Magnetic Fields in O, B, and A Stars

Swetlana Hubrig
Leibniz-Institut für Astrophysik Potsdam (AIP), Potsdam, Germany

Markus Schöller
European Southern Observatory (ESO), Garching, Germany

IOP Publishing, Bristol, UK

ISBN 978-0-7503-2392-5 (ebook)
ISBN 978-0-7503-2390-1 (print)
ISBN 978-0-7503-2393-2 (myPrint)
ISBN 978-0-7503-2391-8 (mobi)

DOI 10.1088/2514-3433/abefcc

Version: 20210601

AAS–IOP Astronomy
ISSN 2514-3433 (online)
ISSN 2515-141X (print)

British Library Cataloguing-in-Publication Data: A catalogue record for this book is available from the British Library.

Published by IOP Publishing, wholly owned by The Institute of Physics, London

IOP Publishing, Temple Circus, Temple Way, Bristol, BS1 6HG, UK

US Office: IOP Publishing, Inc., 190 North Independence Mall West, Suite 601, Philadelphia, PA 19106, USA

To our grandmothers.

Contents

Preface

The last decades have seen a significant step forward in our understanding of the occurrence of magnetic fields in stars with radiative envelopes, in particular in massive stars and in intermediate mass stars at early evolutionary stages. Massive stars determined the early evolution of the universe by driving energetic physical processes that affected the structure of entire galaxies and chemically enriched the interstellar medium. While a number of B-type stars were detected as magnetic already more than half a century back, the first magnetic field detections in O-type stars were achieved less than two decades ago, even though the existence of magnetic O-type stars had been suspected for a long time. The presence of magnetic fields has the potential to change our whole picture about the evolution from O stars via WR stars to supernovae or gamma-ray bursts.

Intermediate mass stars at early evolutionary stages, the pre-main sequence Herbig Ae/Be stars, are surrounded by large amounts of circumstellar gas and dust, making it hard to measure their magnetic fields, which are typically one order of magnitude weaker compared to the fields of their lower-mass pre-main sequence counterpart T Tauri stars. Understanding the interaction between the central stars, their magnetic fields, and their protoplanetary disks is crucial for reconstructing the solar system's history, and to account for the diversity of exo-planetary systems.

Major progress in the studies of stellar magnetism became possible due to the installation of high-resolution spectropolarimeters at 2–4 m class telescopes, such as Narval, ESPaDOnS, and HARPS. Notably, the availability of the FOcal Reducer low dispersion Spectrographs FORS 1 and FORS 2 in spectropolarimetric mode mounted on the 8 m Very Large Telescope allowed astronomers to access massive stars even in galaxies in our neighbourhood.

Studies of stellar magnetic fields present a difficult and challenging part of astrophysics. The aim of this book is to present the current knowledge of stellar magnetism across observational and theoretical studies. We also tried to lay a path that will allow scientists to expand our knowledge of stellar magnetism in the future. Further, challenges resulting from current studies and priorities for future studies are discussed, so as to provide the necessary input for the further progress in the understanding of the magnetic field origin and the role of magnetic fields for the study of stellar evolution. We note that this book represents the authors' perspective, and should not be considered as an exhaustive review covering the full scope of the literature.

A number of magnetic studies described in the book would not have been possible without the invaluable efforts of our colleagues and long-standing collaborators Gautier Mathys, Jorge Federico Gonzalez, Silva P. Järvinen, Ilya Ilyin, Charles R. Cowley, Don W. Kurtz, Alexander F. Kholtygin, S. Drew Chojnowski, Maryline Briquet, Thorsten A. Carroll, Christian A. Hummel, Beate Stelzer, Natalia A. Drake, Zdenek Mikulášek, Manfred Küker. Igor S. Savanov, and Mikhail A. Pogodin, among others.

Particular gratitude is due to Fiorella Castelli for her many years of collaboration. She sadly passed away far too early.

Author Biographies

Swetlana Hubrig

Swetlana Hubrig studied astronomy at the St. Petersburg university and started her professional work at the Zentralinstitut für Astrophysik in Potsdam, where she obtained her PhD in 1989. During the 1990s, she held several positions and grants, both at the Potsdam university and the Astrophysikalisches Institut Potsdam (AIP), habilitating at the university in 2000. From 2001, she worked as operations staff astronomer for the European Southern Observatory on the Very Large Telescope, until she moved back to the Leibniz-Institut für Astrophysik Potsdam to become the head of the stellar physics and stellar activity department in 2009. Swetlana is working on many aspects of the magnetism in stars and is an expert in high-resolution spectroscopy.

Markus Schöller

Markus Schöller studied physics at the universities in Köln and Bonn, where he obtained his PhD in 1996, working at the Max-Planck-Institut für Radioastronomie. Markus joined the European Southern Observatory in 1997, where he mainly works on the Very Large Telescope Interferometer and its scientific instruments. His main science interests are binarity and the magnetism in stars and he specialises in high angular resolution techniques.

Symbols

$\langle B_z \rangle$	mean longitudinal magnetic field
$\langle B \rangle$	mean magnetic field modulus
B_d	dipolar magnetic field
$v \sin i$	projected stellar rotational velocity
T_{eff}	effective temperature
$\log g$	logarithmic surface gravity
S/N	signal-to-noise ratio

Acronyms

AMBER	Astronomical Multi-Beam Recombiner
APOGEE	Apache Point Observatory Galaxy Evolution Experiment
ASAS	All Sky Automated Survey
BOB	B-fields in OB stars
BRITE	BRIght Target Explorer
CFHT	Canada–France–Hawaii-Telescope
CHARA	Center for High Angular Resolution Astronomy
CIR	corotating interaction region
CONICA	COudé Near-Infrared CAmera
CoRoT	Convection, Rotation et Transits planétaire
CPU	central processing unit
CRIRES	CRyogenic high-resolution InfraRed Echelle Spectrograph
cTTS	classical T Tauri star
DAC	discrete absorption component
DI	Doppler Imaging
EPIC	European photon imaging camera
ESO	European Southern Observatory
ESPaDOnS	Echelle SpectroPolarimetric Device for the Observation of Stars
FAP	false alarm probability
FEROS	Fiberfed Extended Range Optical Spectrograph
FOCAS	Faint Object Camera and Spectrograph
FORS	FOcal Reducer and Spectrograph
FUSE	Far Ultraviolet Spectroscopic Explorer
GMOS	Gemini Multi-Object Spectrograph
GRB	gamma-ray burst
HARPS	High Accuracy Radial velocity Planet Searcher
HARPSpol	HARPS polarimeter
HJD	heliocentric Julian Date
HMXB	high-mass X-ray binary
IGR	INTEGRAL Gamma-Ray source
IGW	Internal Gravity Waves
INTEGRAL	INTErnational Gamma-Ray Astrophysics Laboratory
IR	infrared
IRAC	Infrared Array Camera
IRAS	Infrared Astronomical Satellite
IRS	InfraRed Spectrograph
ISM	interstellar medium
IUE	International Ultraviolet Explorer
KELT	Kilodegree Extremely Little Telescope
LAMOST	Large Sky Area Multi-Object Fiber Spectroscopic Telescope
LBT	Large Binocular Telescope
LBV	luminous blue variable
LMC	Large Magellanic cloud
LSD	least-squares deconvolution
LTE	local thermal equilibrium
MAGORI	MAGnetic field ORIgin
MASCARA	Multi-site All-Sky CAmeRA

MCWS	magnetically confined wind shock
MDI	Magnetic Doppler Imaging
MHD	magnetohydrodynamics
MIDAS	Munich Image Data Analysis System
MIDI	MID-infrared Interferometric instrument
MiMeS	Magnetism in Massive Stars
MJD	modified Julian Date
MOST	Microvariability and Oscillations of Stars
MRI	magneto-rotational instability
NACO	NAOS-CONICA
NAOS	Nasmyth Adaptive Optics System
NIR	near-infrared
PACS	Photodetector Array Camera & Spectrometer
PAH	polycyclic aromatic hydrocarbons
PARSEC	PAdova & TRieste Stellar Evolution Code
PEPSI	Potsdam Echelle Polarimetric and Spectroscopic Instrument
PIONIER	Precision Integrated-Optics Near-infrared Imaging ExpeRiment
PIÚ	polarization injection unit for UVES
PMS	pre-main sequence
REE	rare earth element
roAp	rapidly oscillating Ap star
ROSAT	Röntgensatellit
RRM	rigidly-rotating magnetosphere
SALT	Southern African Large Telescope
SB	spectroscopic binary
SDSS	Sloan Digital Sky Survey
SED	spectral energy distribution
SFXT	supergiant fast X-ray transients
sgHMXB	supergiant HMXB
sHR	spectroscopic Hertzsprung–Russell diagram
SLF	stochastic low-frequency variability
SMEI	Solar Mass Ejection Imager
SOFIN	SOviet-FINnish échelle spectrograph
SPB	Slowly pulsating B (star)
SPHERE	SpectroPolarimetric High-contrast Exoplanet REsearch
STELLA	STELLar Activity
STEREO	Solar Terrestrial Relations Observatory
SuperWASP	(Super-) Wide Angle Search for Planets
SVD	Singular Value Decomposition
TAMS	terminal-age main sequence
TBL	Télescope Bernard Lyot
TESS	Transiting Exoplanet Survey Satellite
UV	ultraviolet
UVES	Ultraviolet and Visual Echelle Spectrograph
VALD	Vienna Atomic Line Database
VeSElkA	Vertical Stratification of Element Abundances
VLT	Very Large Telescope
VLTI	Very Large Telescope Interferometer
wTTS	weak-line T Tauri star
WIRE	Wide Field Infrared Explorer

WR	Wolf–Rayet
XMM	X-ray Multi-Mirror Mission
XRF	X-ray flashes
ZAMS	zero-age main sequence
ZCD	Zeeman Component Decomposition
ZDI	Zeeman–Doppler Imaging

Swetlana Hubrig and Markus Schöller

Chapter 1

Introduction

As part of the Henry Draper Memorial classification work at Harvard, Antonia Maury, under the direction of Edward C. Pickering, introduced the designation "peculiar" for the star α^2 CVn, exhibiting anomalous line strengths of certain spectral lines (Maury & Pickering 1897). This discovery, together with further eight stars mentioned to have a peculiar spectrum, defined for the first time the group of chemically peculiar stars. Only half a century later were the abundance peculiarities in a number of main-sequence A- and B-type stars brought into connection with the presence of surface magnetic fields. While a magnetic field in the Sun was already detected in 1908 by Hale (1908), the first detection of a stellar magnetic field was achieved in the A2p star 78 Vir in 1947 by Babcock (1947), who used an observing technique analysing circularly polarised light, similar to Hale.

Observations of 78 Vir began in 1946 April (Babcock 1947) and confirmed Babcock's earlier prediction that magnetic fields of the order of hundreds of Gauss should occur in a number of early-type stars, and that the integrated Zeeman effect resulting from such stellar fields should be observable under certain conditions. A few years later, Babcock (1950) discussed the patchy surface distribution of different chemical elements and large differential velocities between elements, which seemed to be coupled to the magnetic changes. The oblique rotator model, assuming that the magnetic dipole axis is not aligned with the rotation axis, was considered to explain the behaviour of different observables. As the star rotates, the magnetic field and the surface abundance distribution are observed from various aspects, resulting in variability of the measured magnetic field and the spectral line intensities (see also Stibbs 1950). Babcock also reported on the crossover effect that appears as a difference in sharpness of the line profiles of most of the spectral lines, when the right-handed polarised spectrum is compared with the left-handed polarised spectrum. The crossover effect occurs at and near the phases when the mean longitudinal magnetic field phase curve is passing through zero. A suggestion that has been put forward to account for the crossover effect was that the absorption

doi:10.1088/2514-3433/abefccch1

lines may be considered to arise in different surface regions and that these regions are characterised by magnetic fields of different polarity. Thus, different shapes of line profiles result from a combination of Zeeman and Doppler effects.

Follow-up observations of several hundred stars over eleven years were presented by Babcock (1958) in the first catalog of magnetic stars, where he listed 89 stars with definite magnetic field detections, 66 stars with probable detections, 60 stars with sharp lines but no detections and more than 120 stars in which the lines turned out to be too broad to allow any measurement. An important result of his study was that stellar magnetic fields are subject to variations on a timescale of the order of days and with amplitudes of a few hundred to a few thousand Gauss.

Numerous magnetic studies carried out over a few decades after Babcock's first discovery of a magnetic field in an A-type star indicated that the incidence rate of magnetic stars among the intermediate and massive main-sequence stars with spectral types between approximately B2 and F0 is roughly 10%–20% (e.g., Preston 1974) and showed that a majority of the magnetic stars are intrinsically slow rotators (e.g., Slettebak 1954; Abt et al. 1972). The most variable and abnormally intensive lines in the spectra of these stars usually belong to Si, Cr, Mn, Sr, Y, Zr, and the rare earth elements. A division of the chemically peculiar Ap and Bp stars into four groups according to their spectroscopic characteristics was introduced by Preston (1974), who suggested to assign the group CP1 to metallic-line Am/Fm stars, the group CP2 to magnetic Si and SrCrEu stars with late-B and A spectral types, the group CP3 to HgMn stars, and the group CP4 to magnetic He-weak stars. An additional group of chemically peculiar stars assigned to magnetic early B-type stars with anomalously strong He lines was identified by Pedersen & Thomsen (1977).

Abt & Snowden (1973) discussed the frequency of close spectroscopic binaries in stars of different chemical peculiarity and concluded that close-binary formation is inhibited when magnetic fields are present. The authors also suggested that low rotational velocities of magnetic stars are due to magnetic braking and not to tidal interaction. Michaud (1970) considered diffusion in the atmospheres of chemically peculiar stars and was able to account for the overabundance of Si, Mn, and heavy elements and deficiencies of helium and oxygen in the temperature regions where these stars are observed, assuming that the atmospheric turbulence is sufficiently low.

In particular the last decades have seen a significant step forward in the understanding of the occurrence of magnetic fields in stars of different mass and different age. Studies of several integral magnetic observables provided by the moment technique introduced by Mathys (1988) revealed that the geometrical structures of the magnetic fields of Ap and late-type Bp stars appear in general to depart only slightly from centred dipoles (e.g., Mathys & Hubrig 1997; Mathys 2017). The Doppler Imaging code initially developed by Goncharskij et al. (1982) was extended to a code capable of simultaneously reconstructing the magnetic field vector and the chemical element distribution over the stellar surfaces of Ap and late-type Bp stars (e.g., Khokhlova et al. 1997; Piskunov & Kochukhov 2002). Successive studies using full Stokes vector observations confirmed that magnetic observables can be reasonably well reproduced by a dipolar model (e.g., Rusomarov et al. 2013). Rapidly oscillating Ap (roAp) stars, constituting a subset of cool Ap stars, were discovered by Kurtz (1982). They pulsate in high-overtone non-radial p-modes with periods in the range of 5–25 min and offer a unique opportunity to observe the interaction of p-modes with strong magnetic fields. Previously, this had only been achieved for the Sun. The unique properties of the roAp stars have the potential to allow 3D mapping of their pulsational modes, magnetic field geometries and abundance distributions within the observable atmospheric layers (see Chapter 4).

Significant progress has been achieved in studies of the occurrence of magnetic fields, their strengths and their geometries in massive O and B stars. Massive early-B stars and O stars dominate the ecology of the universe as driving agent, through their luminosities and mechanical inputs, such as winds and supernova explosions. They are important sources of ionising photons and elements heavier than helium, produced through nuclear burning, and determined the early evolution of the universe. Recent surveys of the magnetism in massive stars aimed at studies of the interaction between the magnetic fields and the stellar winds, the evolution of the fields themselves and the impact of the fields on the stellar evolution (see Chapter 5). As of today, about a dozen O stars and several dozen early B-type stars are known to possess oblique, predominantly dipolar magnetic fields with surface field strengths in the range from a few Gauss up to more than ten kG (e.g., Grunhut et al. 2017; Schöller et al. 2017).

Magnetic fields are also of utmost importance in the evolution of the stellar nurseries, the interstellar molecular clouds, and they are present at all stages of stellar evolution. The magnetic field is omnipresent at all scales and stages of star formation and thus drives several key processes during the formation of stars, such as removing angular momentum during the main accretion phase, affecting disk initial conditions and subsequent planet formation, and allowing late-stage accretion onto pre-main-sequence stars. Systematic magnetic studies of intermediate-mass pre-main-sequence Herbig Ae/Be stars indicated dipole-like field configurations, similar to those observed in magnetic OBA stars on the main-sequence (e.g., Hubrig et al. 2011). However, their longitudinal magnetic field strengths are in general much lower, usually of the order of tens of Gauss (e.g., Järvinen et al. 2019). So far, only about 20 Herbig Ae/Be stars were reported to host magnetic fields and the magnetic field geometry was constrained only for 2 Herbig Ae/Be stars (e.g., Hubrig et al. 2015). Obviously, studies of the magnetic field topology in these stars are extremely important because they will enable researchers to improve their insight into how the magnetic fields are generated during the pre-main-sequence stage and how they interact with their environment, including their impact on the planet formation processes and the planet–disk interaction. Understanding the interaction between the magnetic fields of central stars and their protoplanetary disks is also crucial for recent efforts to account for the diversity of exo-planetary systems (see Chapter 6).

The notable advances in the studies of stellar magnetism reported in this book became possible through the significant progress in spectropolarimetric surveys of surface magnetic fields, which made frequent use of the spectropolarimeters ESPaDOnS (Echelle SpectroPolarimetric Device for the Observation of Stars) at the Canada–France–Hawaii telescope, NARVAL at the Bernard Lyot 2 m Cassegrain telescope, the Main Stellar Spectrograph of the Russian 6 m telescope, FORS 1 and FORS 2 (FOcal Reducer low dispersion Spectrographs) at the ESO Very Large Telescope, and HARPSpol (High Accuracy Radial velocity Planet Searcher) at ESO/La Silla. Furthermore, more sophisticated observational techniques and analysis tools (see Chapters 2 and 3) have been used to probe magnetic fields in different types of stars, including stars and systems not systematically studied with spectropolarimetry in the past, such as metallic-line Am stars, HgMn stars, β Cephei stars, slowly pulsating B stars, Be stars, Wolf–Rayet stars, and high-mass X-ray binaries (see Chapter 7).

References

Abt, H. A., & Snowden, M. S. 1973, ApJS, 25, 137
Abt, H. A., Chaffee, F. H., & Suffolk, G. 1972, ApJ, 175, 779
Babcock, H. W. 1947, PhRv, 72, 83
Babcock, H. W. 1950, Natur, 166, 249
Babcock, H. W. 1958, ApJS, 3, 141
Goncharskij, A. V., Stepanov, V. V., Khokhlova, V. L., & Yagola, A. G. 1982, AZh, 59, 1146
Grunhut, J. H., Wade, G. A., Neiner, C., et al. 2017, MNRAS, 465, 2432
Hale, G. E. 1908, ApJ, 28, 315
Hubrig, S., Mikulášek, Z., González, J. F., et al. 2011, A&A, 525, L4
Hubrig, S., Carroll, T. A., Scholler, M., & Ilyin, I. 2015, MNRAS, 449, L118
Järvinen, S. P., Carroll, T. A., Hubrig, S., Ilyin, I., & Schöller, M. 2019, MNRAS, 489, 886
Khokhlova, V. L., Vasilchenko, D. V., Stepanov, V. V., & Tsymbal, V. V. 1997, AstL, 23, 465
Kurtz, D. W. 1982, MNRAS, 200, 807
Mathys, G., & Hubrig, S. 1997, A&AS, 124, 475
Mathys, G. 1988, A&A, 189, 179
Mathys, G. 2017, A&A, 601, A14
Maury, A. C., & Pickering, E. C. 1897, AnHar, 28, 1
Michaud, G. 1970, ApJ, 160, 641
Pedersen, H., & Thomsen, B. 1977, A&AS, 30, 11
Piskunov, N., & Kochukhov, O. 2002, A&A, 381, 736
Preston, G. W. 1974, ARA&A, 12, 257
Rusomarov, N., Kochukhov, O., Piskunov, N., et al. 2013, A&A, 558, A8
Schöller, M., Hubrig, S., Fossati, L., et al. 2017, A&A, 599, A66
Slettebak, A. 1954, ApJ, 119, 146
Stibbs, D. W. N. 1950, MNRAS, 110, 395

Chapter 2

Basics of Stellar Magnetic Field Studies

As of today, all magnetic field diagnostics applied in the studies of stars with radiative envelopes are based on the Zeeman effect. In this chapter, the concepts of the Stokes formalism and the oblique rotator are introduced, followed by explanations how the Zeeman effect influences the spectral appearance in a magnetic field and how this can be used to determine magnetic field strengths and geometries with different analytical methods.

2.1 Introduction to the Field

A magnetic field can be described as a vector $\vec{B}(\vec{x}, t)$ for every point in space and time, i.e., as $(B_x, B_y, B_z)(x, y, z, t)$. For a star, only part of this field can be accessed, usually close to its surface. Further, one cannot retrieve the full vector, but only some aspects of it, like the mean magnetic field modulus $\langle B \rangle$ or the mean longitudinal magnetic field $\langle B_z \rangle$. Observations are averaged over the part of the star that is visible and weighted by the local emergent line intensity of the lines used in the measurements. Depending on the element employed, the magnetic field can be measured in the star in different depths of its atmosphere, different heights above its surface, or different locations on the surface, depending on the element distribution over the star. The field itself may vary with time, possibly through dynamic effects in the stellar atmosphere, but mostly due to stellar rotation. Accordingly, a specific, complex, and fundamental aspect of stellar magnetic field studies is to find ways of extracting physically meaningful constraints for the field from disk-integrated observational data. Especially the aspect-sensitive mean longitudinal magnetic field $\langle B_z \rangle$ is frequently used to detect magnetic fields and represents a good tracer to determine the geometry of a magnetic field and the rotation period of a star. The mean longitudinal magnetic field $\langle B_z \rangle$ is a line-intensity weighted average over the visible hemisphere of a star of the magnetic field component directed along the line of sight.

doi:10.1088/2514-3433/abefccch2

Other observational and analytical methods to diagnose the presence and structure of magnetic fields in stellar atmospheres include broad-band linear polarisation, the moment technique, and Zeeman–Doppler imaging. The Hanle effect (Hanle 1923), which concerns the modification of the polarising effects of resonance-line scattering by a magnetic field, can in principle be used as a diagnostic of stellar magnetic fields. It has been almost exclusively applied in the solar context, to diagnose magnetic fields in solar prominences and the upper solar atmosphere and as a diagnostic of solar turbulent field strengths (e.g., Stenflo 1982). According to Ignace et al. (1997), the best potential targets to use the Hanle effect as a diagnostic of magnetic fields are O- and B-type stars with resonance lines of highly ionised species, such as C IV or Si IV, among others.

Importantly, physically realistic models of the geometric structure of magnetic fields in stellar atmospheres can be achieved only if linear polarisation measurements are implemented, i.e., the most complete way to study stellar magnetic fields is to observe both linear and circular polarisation. However, linear polarisation is a second-order effect in the magnetic field and, in addition, in disk-integrated observations mutual cancellation of the contributions of different regions on the stellar surface is much more effective for linear polarisation than for circular polarisation.

2.2 The Stokes Formalism

Spectropolarimeters make use of polarisation analysing optics, typically sets of half- and quarter-waveplates together with a Wollaston prism, a Fresnel rhomb, or another type of prism. They allow to obtain linearly and circularly polarised spectra. Quarter-waveplates are used to convert circularly polarised light into linearly polarised light, half-waveplates manipulate the polarisation axes, and a Wollaston prism splits the light into two beams with orthogonal linear polarisation. The waveplates are typically rotated to obtain different orientations of the polarisation and to calibrate instrumental effects.

The resulting spectra are linearly polarised for different waveplate angles, e.g., I_0, I_{45}, I_{90}, I_{135}, or right- and left-handed circularly polarised, i.e., I_r and I_l. They are usually treated together in the Stokes formalism (Stokes 1851), which combines for every (spatial or, in the cases described in this book, spectral) element the total flux (I) with two linear polarisations (Q and U) and circular polarisation (V): $\vec{S} = (I, Q, U, V)$.

These four values are calculated from the obtained data in the following way:

$$I = I_0 + I_{90} \tag{2.1}$$

$$\text{or} \quad I = I_{45} + I_{135} \tag{2.2}$$

$$\text{or} \quad I = I_r + I_l \tag{2.3}$$

$$Q = I_0 - I_{90} \tag{2.4}$$

$$U = I_{45} - I_{135} \tag{2.5}$$

$$V = I_r - I_l. \tag{2.6}$$

The degree of linear polarisation is calculated as

$$P = \frac{\sqrt{Q^2 + U^2}}{I} \tag{2.7}$$

and the polarisation angle as

$$\phi = \frac{1}{2} \arctan \frac{U}{Q}. \tag{2.8}$$

The Stokes formalism is especially suited for astronomical data that do not have phase information. For data with phase information, the Jones formalism (e.g., Jones 1941) is more adequate, as it works on the electromagnetic waves and not on intensities. The propagation of polarised light through optical systems can be described by Müller matrices (Müller 1943), which have a size of 4×4 and are real.

In the following, mainly the use of the Stokes I and the Stokes V spectra or their fraction V/I is illustrated.

2.3 The Oblique Rotator Model

When observing stars over several epochs, it is quite common to see variations of the strength of their magnetic field, especially for the longitudinal magnetic field. Most of these variations can easily be described by the oblique rotator model, which was first introduced by Stibbs (1950). The stellar magnetic field is rotating with the star and one can see different aspects of it.

The oblique rotator assumes a dipolar magnetic field tilted with the obliquity angle β with respect to the rotation axis, which itself is inclined with respect to the line of sight by the angle i. In the case that i is zero, one is looking at the rotation pole and no modulation of the magnetic field with rotation can be seen. The same is true in the case when β is zero, i.e., the magnetic field axis is aligned with the rotation axis.

A general description for the strength of the observed longitudinal magnetic field was presented by Preston (1967):

$$\langle B_z \rangle = \frac{1}{20} \frac{15 + u}{3 - u} B_d(\cos\beta \cos i + \sin\beta \sin i \cos 2\pi t/P) \tag{2.9}$$

with $\langle B_z \rangle$ the longitudinal magnetic field, β the angle between the rotation axis and the magnetic axis, i the angle between the rotation axis and the line of sight, P the rotation period of the star, u the limb-darkening coefficient, and B_d the strength of the dipolar magnetic field.

One can easily see that in this formalism the two angles i and β can be exchanged for each other. As a consequence, one can determine two angles, but cannot distinguish which one is which. To obtain the angles, the fraction (including the sign) of the minimum $\langle B_z \rangle_{\min}$ and the maximum $\langle B_z \rangle_{\max}$ longitudinal magnetic field values is calculated:

$$r = \frac{\langle B_z \rangle_{\min}}{\langle B_z \rangle_{\max}} \tag{2.10}$$

$$= \frac{\cos\beta\cos i - \sin\beta\sin i}{\cos\beta\cos i + \sin\beta\sin i} \tag{2.11}$$

and β depending on i (or i depending on β) can be calculated:

$$\beta = \arctan\left(\frac{1-r}{1+r}\cot i\right). \tag{2.12}$$

It is of course possible to deduce the inclination i from other sources. If the rotation period P is known, e.g., through measurements of the longitudinal magnetic field over the full rotation period, as well as the projected rotation velocity $v \sin i$ and the stellar radius R, then one can calculate i through (Preston 1969):

$$\sin i = \frac{v \sin i}{v_{\mathrm{eq}}} = \frac{v \sin i \; P}{50.6\; R} \tag{2.13}$$

with v_{eq} the rotation velocity at the stellar equator in km s^{-1}, P in days and R in solar radii. Alternatively one might use inclination information from a surrounding disk or the inclination of a close binary orbit.

The geometric dependence of the longitudinal magnetic field on the obliquity angle β and the inclination angle i is illustrated in Figure 2.1.

2.4 The Zeeman Effect

When an atom is located in a magnetic field, its energy levels are modified and consequently also its spectral lines. The lines may split into several components and their polarisation properties are altered. Zeeman (1897) was the first to detect this effect in the laboratory.

A magnetic field alters the Hamiltonian in the Schrödinger equation and adds two new terms:

$$H_B = \frac{eh}{4\pi mc}\left(\vec{L} + 2\vec{S}\right)\vec{B} + \frac{e^2}{8mc^2}\left(\vec{B} \times \vec{r}\right)^2 \tag{2.14}$$

with e the charge of the electron, h the Planck constant, m the mass of the electron, c the speed of light, $\vec{L}$ the total angular orbital momentum, $\vec{S}$ the total spin of the electron cloud, $\vec{B}$ the magnetic field vector, and $\vec{r}$ the sum over all positions of the electrons with respect to the nucleus.

The first term of the Hamiltonian describes the linear Zeeman effect, the second term the quadratic Zeeman effect and is also called the diamagnetic term. All objects studied in this book typically have magnetic fields low enough that the quadratic Zeeman effect can be neglected. In this case, the Hamiltonian can be rewritten as

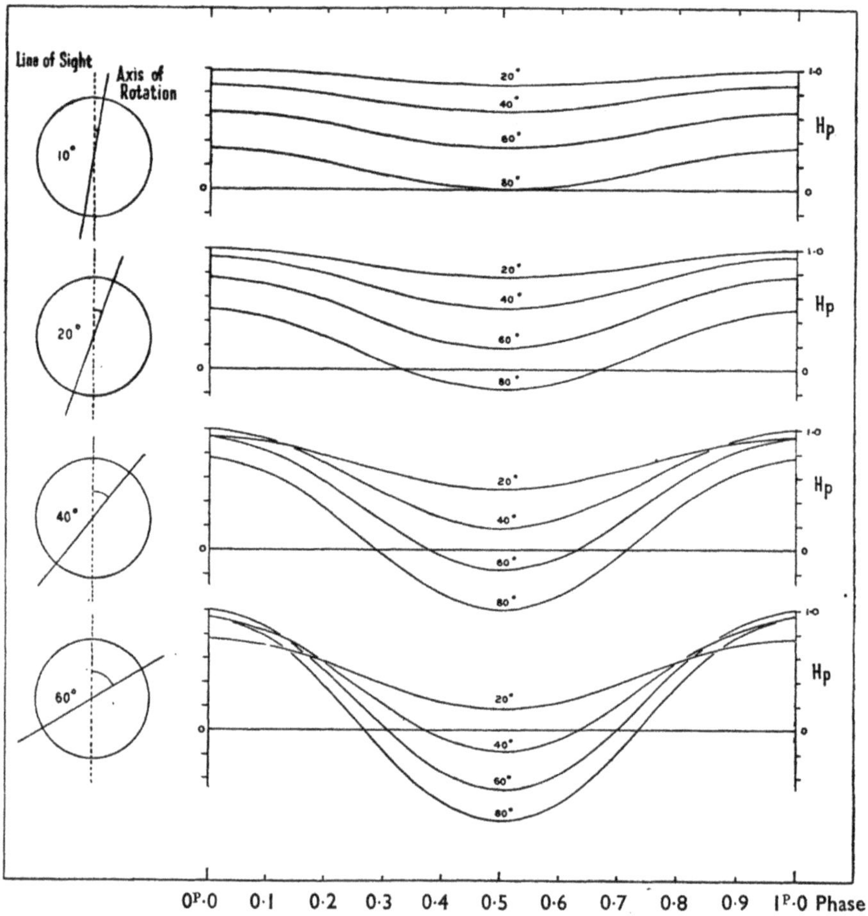

Figure 2.1. Apparent variations of the longitudinal magnetic field $\langle B_z \rangle$ during the rotation of a star when viewed obliquely to the axis of rotation, the magnetic and rotation axes not being coincident. The four curves shown for each inclination to the line of sight are for obliquity angles 20°, 40°, 60°, and 80° of the magnetic pole. Please note that the H_p used in this figure is in fact the longitudinal magnetic field $\langle B_z \rangle$. Credit: Figure 1 from Stibbs (1950).

$$H_B = \mu_0\left(\vec{L} + 2\vec{S}\right)\vec{B} \tag{2.15}$$

with $\mu_0 = eh/4\pi mc = 9.27 \times 10^{-21}$ erg G^{-1} the Bohr magneton.

For weak magnetic fields, where H_B can be treated as a perturbation, the atomic energy levels split up into $2J + 1$ sublevels, with J the total angular momentum quantum number. The difference in energy between a sublevel and the magnetically undisturbed level is given by:

$$\Delta E = \mu_0 gBM, \tag{2.16}$$

where g is the Landé factor and $M = (-J, -J + 1, ..., 0, ..., J - 1, J)$ the magnetic quantum number, the component of the total angular momentum in the direction of the magnetic field with strength B.

For light elements, typically up to the iron group, and magnetic fields lower than about 1 MG, $\vec{L}$ and $\vec{S}$ are coupled, i.e., the individual electron spins and the individual angular orbital momenta add up separately, before they add up to the total angular momentum $\vec{J}$:

$$\vec{J} = \vec{L} + \vec{S}. \tag{2.17}$$

This is called spin–orbit coupling or Russell–Saunders coupling. For heavier elements or stronger magnetic fields, $\vec{L}$ and $\vec{S}$ decouple and the energy levels in the atom change dramatically. This is called the Paschen–Back effect. While for the stellar objects described in this book the Paschen–Back effect does not play a role, the partial Paschen–Back effect has to be taken into account for the strongest magnetic fields found in Ap/Bp stars, as well as for certain lines already for magnetic fields of a few kG. For example, Mathys (1990) found that the two lines Fe II λ6147.7 and Fe II λ6149.2 were splitting asymmetrically, with one of the σ components deeper and narrower than the other.

Under the assumption of Russell–Saunders coupling and in a regime where the Paschen–Back effect is negligible one can derive the following equation for the Landé factor g and the quantum numbers L, S, and J (e.g., Beckers 1969 and references therein):

$$g = 1 + \frac{J(J + 1) + S(S + 1) - L(L + 1)}{2J(J + 1)}. \tag{2.18}$$

The effective Landé factor $g_{\rm eff}$ of a given transition can be calculated using the total angular momentum quantum numbers J and Landé factors g of the lower level (l) and the upper level (u) as follows:

$$g_{\rm eff} = \frac{1}{2}(g_{\rm l} + g_{\rm u}) + \frac{1}{4}(g_{\rm l} - g_{\rm u})[J_{\rm l}(J_{\rm l} + 1) - J_{\rm u}(J_{\rm u} + 1)], \tag{2.19}$$

where in the case of $J_l = J_u$, $g_{\rm eff}$ is simply the average of the Landé factors of the lower and upper levels.

Dipole transitions between different energy levels in an atom are only allowed, if the difference ΔM between the magnetic quantum numbers of the two states are −1, 0, or +1. Transitions resulting from $\Delta M = 0$ are called the π components and transitions with $\Delta M = \pm 1$ the σ components. The π components are located symmetrically around the position of the magnetically undisturbed line. For every σ component corresponding to a transition with $\Delta M = -1$, there is a symmetrical σ component corresponding to a transition with $\Delta M = +1$. Light coming from a π component is linearly polarised, while light coming from the σ components is circularly polarised. As light from a π component is not radiating into the direction of the magnetic field, the net flux of such a component is zero when looking directly on the magnetic pole. Contrary flux from a σ component is zero when looking at a magnetic equator.

Under the assumption of Russell–Saunders coupling and negligible Paschen–Back effect, further equations can be derived (e.g., Beckers 1969 and references therein).

For the splitting of the components g_π, $g_{\sigma+}$, and $g_{\sigma-}$ one obtains:

$$g_\pi = M_u(g_u - g_l) \tag{2.20}$$

$$g_{\sigma+} = M_u g_u - M_l g_l, \quad \text{with} \quad M_l = M_u + 1 \tag{2.21}$$

$$g_{\sigma-} = M_u g_u - M_l g_l, \quad \text{with} \quad M_l = M_u - 1 \tag{2.22}$$

with the indices π, $\sigma+$, and $\sigma-$ indicating the π, right-hand σ, and left-hand σ components. u and l designate the upper and lower levels and M is the magnetic quantum number.

The intensities I_π, $I_{\sigma+}$, and $I_{\sigma-}$ of the components for an optically thin emitting gas viewed at right angles to the magnetic field are:

$$[J_u = J_l + 1]\colon\ I_\pi = J_u^2 - M_u^2 \tag{2.23}$$

$$I_{\sigma+} = \frac{1}{4}(J_u - M_u)(J_u - M_u - 1) \tag{2.24}$$

$$I_{\sigma-} = \frac{1}{4}(J_u + M_u)(J_u + M_u - 1) \tag{2.25}$$

$$[J_u = J_l]\colon\ I_\pi = M_u^2 \tag{2.26}$$

$$I_{\sigma+} = \frac{1}{4}(J_u - M_u)(J_u + M_u + 1) \tag{2.27}$$

$$I_{\sigma-} = \frac{1}{4}(J_u + M_u)(J_u - M_u + 1) \tag{2.28}$$

$$[J_u = J_l - 1]\colon\ I_\pi = (J_u + 1)^2 - M_u^2 \tag{2.29}$$

$$I_{\sigma+} = \frac{1}{4}(J_u + M_u + 1)(J_u + M_u + 2) \tag{2.30}$$

$$I_{\sigma-} = \frac{1}{4}(J_u - M_u + 1)(J_u - M_u + 2). \tag{2.31}$$

The intensities $\tilde{I}_\pi$, $\tilde{I}_{\sigma+}$, and $\tilde{I}_{\sigma-}$ for an arbitrary angle ϕ between the field direction and the line of sight are:

$$\tilde{I}_\pi = I_\pi \sin^2 \phi \tag{2.32}$$

$$\tilde{I}_{\sigma+} = I_{\sigma+}(1 \pm \cos\phi)^2 \tag{2.33}$$

$$\tilde{I}_{\sigma-} = I_{\sigma-}(1 \mp \cos\phi)^2, \tag{2.34}$$

where the $\pm$ respective $\mp$ are applicable for right- and left-handed circular analysers.

Historically, the split of a spectral line into three different components (a Zeeman triplet) was well understood and called the "normal" Zeeman effect. A split into a different number of components was called the "anomalous" Zeeman effect. An overview about all the existing six different types of Zeeman patterns can be found in Figure 2.2, which was originally published in Mathys (1989). As can be seen, the Fe I and Fe II transitions alone cover already all different pattern types. Notably, because the rare earth element (REE) lines often arise from transitions between atomic levels of high angular momentum J, they tend to have very broad anomalous Zeeman patterns. Among the 112 REE lines that were analysed by Järvinen et al. (2021), only four had a "normal" triplet pattern. Effective Landé factors for REE are on average smaller and span a narrower range of values than those of the lines of Fe and Cr. Statistically, the studied REE transitions display all types of "anomalous" Zeeman patterns, with no particular preference overall. The diversity of the Zeeman patterns is due to a complex combination that is additionally compounded by the effect of the hyperfine structure, which affects several of the REE at various degrees. This effect may in particular be responsible for the frequently observed departures from symmetry of the Stokes I profiles of some REE spectral lines, which form in a regime of Paschen–Back effect rather than Zeeman effect (e.g., Landolfi et al. 2001).

Several magnetic slowly rotating Ap stars show the strong Lu II λ5984 line with a very broad pattern of hyperfine structure. In the presence of a magnetic field, the effect of the hyperfine structure can be remarkable in changing the Zeeman pattern, especially when the Zeeman splitting is comparable to the hyperfine structure splitting (Hubrig et al. 2002). The results of modelling the Lu II λ5984 line accounting for the Paschen–Back effect on the hyperfine structure are shown in Figure 2.3. The line profiles were computed with a radiative code that makes use of the Milne–Eddington approximation. The computations show that with an increasing strength of the magnetic field, the line profile of the Lu II λ5984 line becomes shallower and broader. The calculations performed for this line in the spectrum of Przybylski's star, assuming the appropriate atmospheric parameters, are presented in Figure 2.4.

2.5 Broad-band Linear Polarisation

Observations of broad-band linear polarisation, which is caused by different saturation levels of the π and σ components of a spectral line in the presence of a magnetic field, were reported in the past for a few chemically peculiar magnetic Ap stars (e.g., Leroy 1989; Landolfi et al. 1993). This differential effect for Zeeman components is qualitatively similar for all lines, so that in broad-band observations the contributions of all lines add up. Monitoring of broad-band linear polarisation of a few Ap stars over a few years was used to establish the presence of their

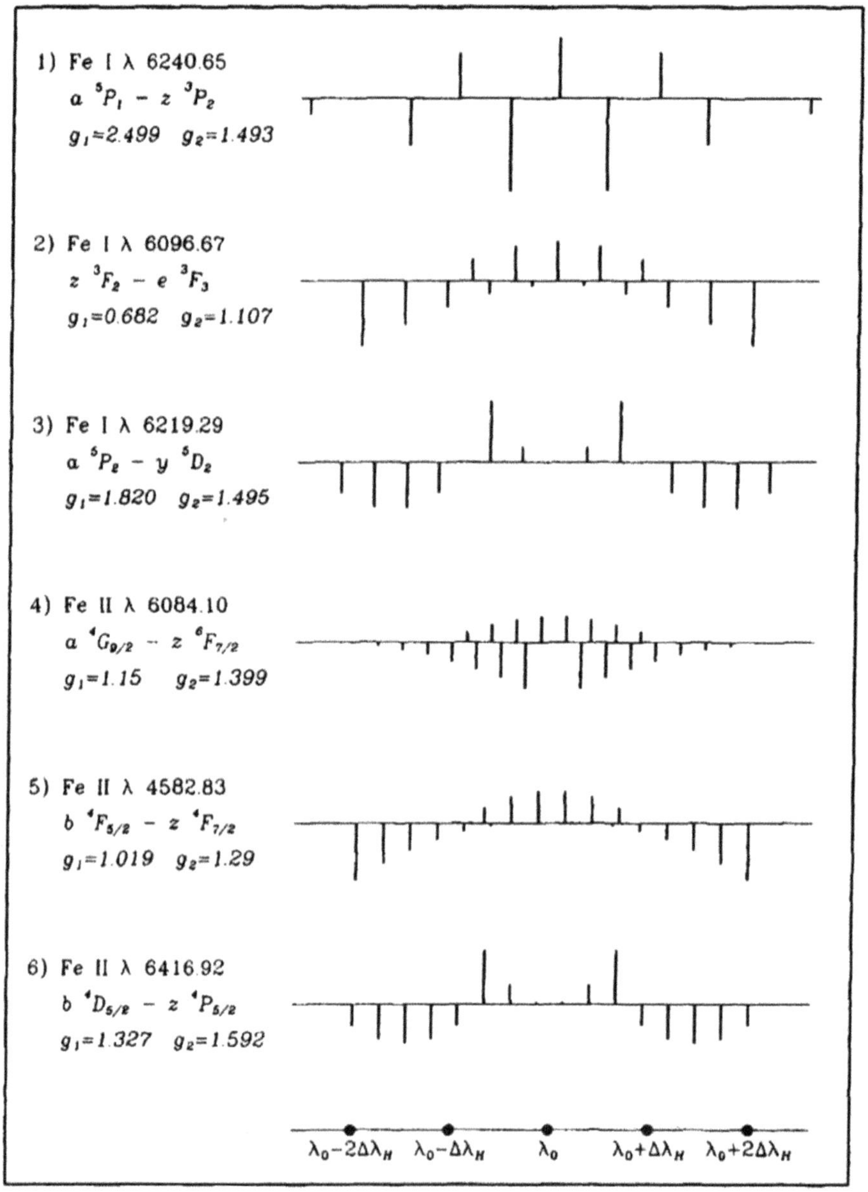

Figure 2.2. The six types of Zeeman pattern. Each Zeeman component is represented by a vertical bar whose length is proportional to its relative strength. The π components appear above the horizontal (wavelength) axis, and the $\sigma_\pm$ components below it. The unit length along the wavelength axis is $\Delta\lambda_H$, as represented at the bottom of the figure. Transitions corresponding to lines actually used to diagnose magnetic fields in stellar atmospheres are represented. Note: the weakest σ components of the patterns of types 4 and 5, and the weakest π components of the pattern of type 6, which have a very small relative strength, may not be easily distinguished on the figure, but are definitely present. In contrast, the central π component of the pattern of type 3 is forbidden. Credit: Mathys (1989).

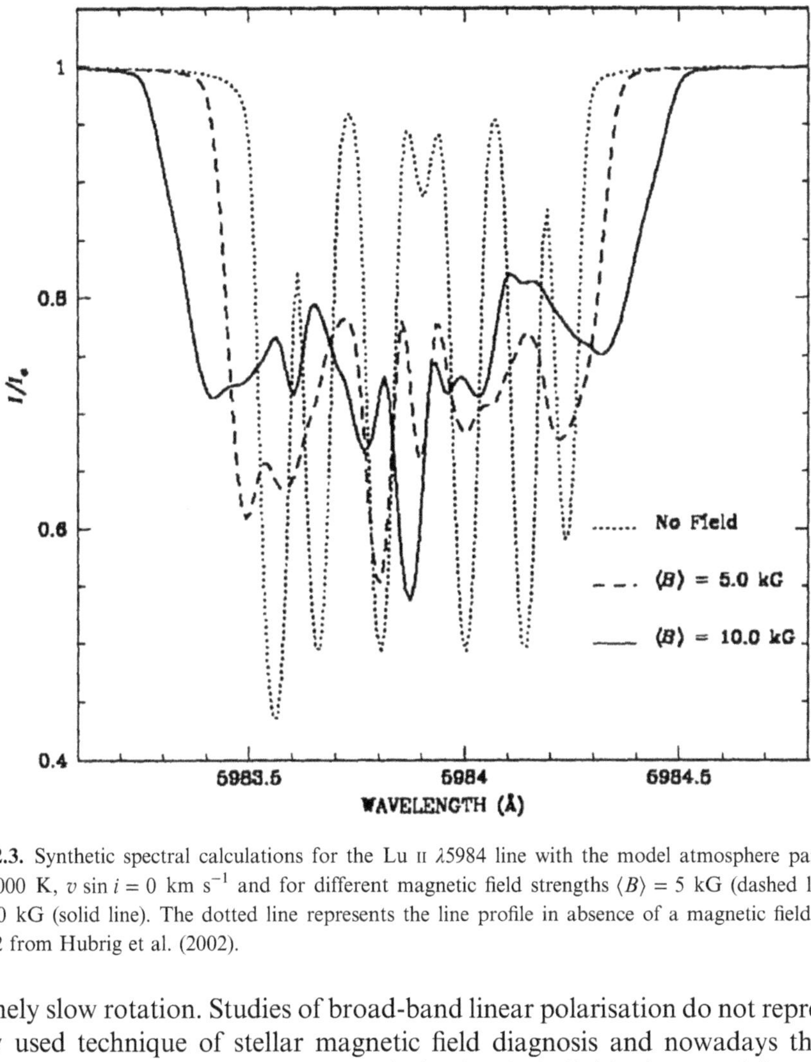

Figure 2.3. Synthetic spectral calculations for the Lu II λ5984 line with the model atmosphere parameters $T_{\rm eff} = 7000$ K, $v \sin i = 0$ km s^{-1} and for different magnetic field strengths $\langle B \rangle = 5$ kG (dashed line) and $\langle B \rangle = 10$ kG (solid line). The dotted line represents the line profile in absence of a magnetic field. Credit: Figure 2 from Hubrig et al. (2002).

extremely slow rotation. Studies of broad-band linear polarisation do not represent a widely used technique of stellar magnetic field diagnosis and nowadays they are superseded by observations of linear polarisation in individual spectral lines.

2.6 Magnetic Field Modulus

In spectra of stars that have sufficiently small projected rotational velocities and magnetic fields strong enough to exceed the rotational Doppler broadening, certain lines with suitable Zeeman patterns show resolved Zeeman split lines in unpolarised light (Preston 1971; Mathys 1990). These lines can be used to estimate the line-intensity weighted average of the modulus of the magnetic field over the visible stellar hemisphere, the mean magnetic field modulus $\langle B \rangle$. Although to some degree dependent on the geometry of the observation, $\langle B \rangle$ has the advantage of being primarily determined by the intrinsic stellar magnetic field strength. An example of a suitable split line is shown in Figure 2.5 (Mathys et al. 1997).

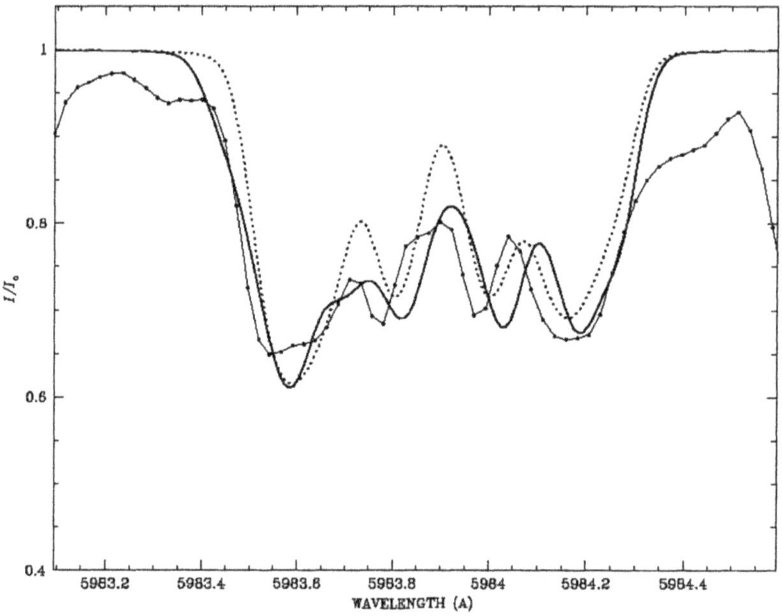

Figure 2.4. Comparison between the synthetic spectrum and observations of the Lu II λ5984 line in the spectrum of Przybylski's star. The synthetic spectrum is calculated with the atmospheric parameters $T_{\rm eff}$ = 6600 K, $v \sin i$ = 3.5 km s^{-1}, and $\langle B \rangle$ = 3.0 kG (solid line). The dotted line represents calculations without a magnetic field. Observations are indicated by dots connected with a solid line. Credit: Figure 3 from Hubrig et al. (2002).

Using the wavelength separation of the magnetically split components in a given absorption line and knowing the Landé factors g of the associated energy levels (see Equations (2.18) and (2.19) in Section 2.4 for the calculation of g and $g_{\rm eff}$), it is straightforward to compute the mean magnetic field modulus $\langle B \rangle$ in units of Gauss according to

$$\langle B \rangle = \frac{\Delta\lambda}{k\, g_{\rm eff}\, \lambda_0^2}. \tag{2.35}$$

Here, $\Delta\lambda$ is the separation in Å between an outer and the central split component or half of the separation between the outer components in a triplet profile, λ_0 is the rest wavelength of the line in Å and $k = 4.67 \times 10^{-13}$ Å^{-1} G^{-1} is a constant.

Spectroscopy at longer wavelengths provides several advantages in terms of resolving magnetically split lines. The wavelength separation of magnetically split components is proportional to the square of the wavelength, while the Doppler effect has a linear dependence on wavelength. Hence, for a fixed effective Landé factor, the spectral resolution required to resolve the split components of a line can be significantly lower in the near-infrared than in the optical.

As stated above, a suitable Zeeman pattern is needed to carry out measurements of magnetic field moduli. Mathys et al. (1997) and Mathys (2017) both chose to use the Fe II λ6149.2 line, which is a true doublet, with the σ and π components showing

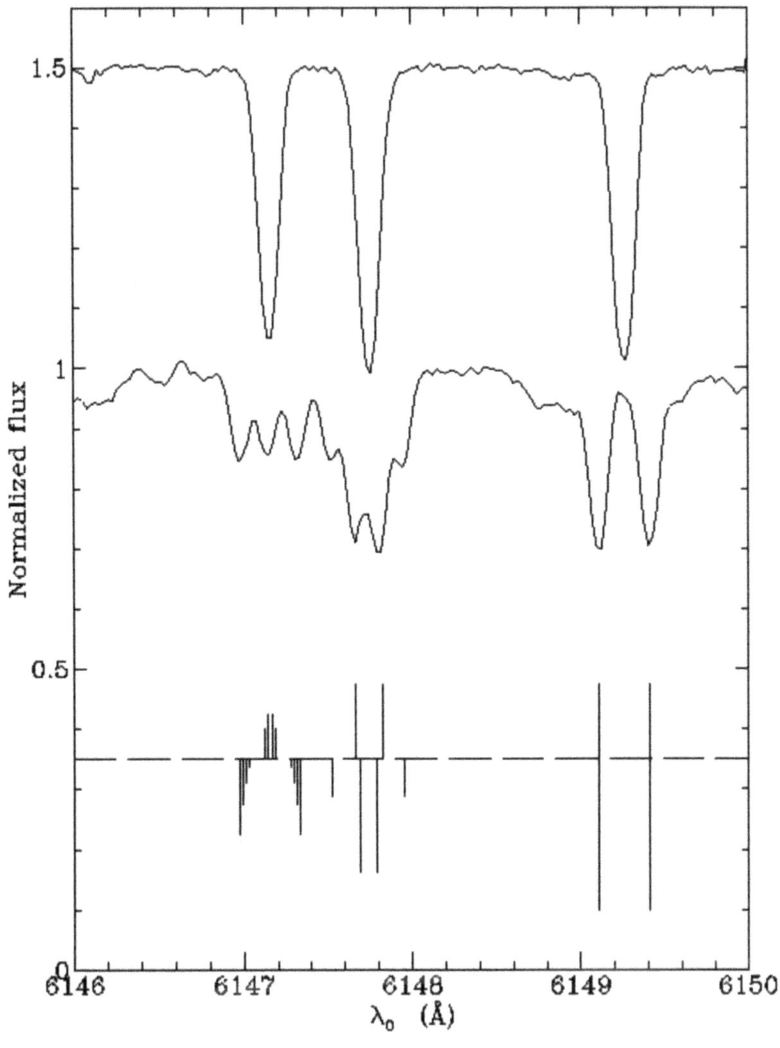

Figure 2.5. Portion of the spectra of HD 133792 (shifted in intensity by 0.5; unresolved lines) and of HD 94660 containing the lines Cr II λ6147.1, Fe II λ6147.7 and Fe II λ6149.2, and Zeeman patterns of those lines. The patterns are represented in the conventional manner, by bars whose length is proportional to the relative strength of the components. The π components appear above the horizontal (wavelength) axis, the σ components below it. For the sake of clarity, the wavelengths in the stellar spectra have been reduced to the laboratory reference frame. Credit: Mathys et al. (1997), reproduced with permission © ESO.

the same displacement, as can be seen in Figure 2.5. The other two lines shown in Figure 2.5 do not show such a simple structure and are more difficult to treat. A number of lines appearing magnetically split in strongly magnetic stars resemble pseudo doublets, e.g., the C IV λ5801.31 line in the spectra of magnetic O-type stars. Treating such lines like a true doublet will lead to an underestimation of $\langle B \rangle$, as the actual component spacing is expected to reduce the separation of the red and blue components of the observed pseudo doublet (e.g., Järvinen et al. 2021).

2.7 Autocorrelation

Another technique to detect surface magnetic fields from Stokes I spectra was suggested by Borra & Deschatelets (2015). The autocorrelation of the integral light spectrum is used to measure the width of spectral lines and to disentangle the magnetic broadening (which scales with λ^2) from other broadening effects (which scale with λ), using their different wavelength dependence. This technique can also be applied to low-resolution spectra and be utilised on stars that are too faint for high-resolution observations even with large telescopes. It is especially intriguing that the autocorrelation technique can be applied to readily available data archives from numerous spectrographs all over the world to obtain reliable statistics on the occurrence of magnetic fields in stars in clusters and the field at different evolutionary stages, including the neighbouring galaxies.

2.8 Spectral Line Moments in Stokes I and Stokes V

A further method to overcome rotational broadening is the moment technique. It allows the determination of the crossover effect and of the mean quadratic field (e.g., Mathys 1993; Mathys & Hubrig 2006). The profile shapes of spectral lines observed in Stokes I and Stokes V are characterised by their moments of various orders about the line centre and can be interpreted in terms of quantities related to the magnetic field. The analysis is usually based on the consideration of samples of reasonably unblended lines and critically depends on the number of lines that can be employed.

The mean quadratic magnetic field is defined as

$$\langle B_q \rangle = \left(\langle B^2 \rangle + \langle B_z^2 \rangle \right)^{1/2}. \tag{2.36}$$

Here, $\langle B^2 \rangle$ is the mean square magnetic field modulus, i.e., the average over the stellar disc of the square of the modulus of the magnetic field vector, weighted by the local emergent line intensity, while $\langle B_z^2 \rangle$ is the mean square longitudinal magnetic field, i.e., the average over the stellar disc of the square of the line-of-sight component of the magnetic vector, weighted by the local emergent line intensity.

The mean quadratic magnetic field is determined from the study of the second-order moments of the line profiles recorded in unpolarised light (i.e., in Stokes I). The second-order moment $R_I^{(2)}(\lambda_I)$ of a spectral line profile recorded in unpolarised light about its centre of gravity λ_I is defined as

$$R_I^{(2)}(\lambda_0) = \frac{1}{W_\lambda} \int r_{\mathcal{F}_I}(\lambda - \lambda_0)(\lambda - \lambda_0)^2 \, d\lambda. \tag{2.37}$$

The integration runs over the whole width of the observed line (see Mathys 1988 for details). W_λ is the line equivalent width; $r_{\mathcal{F}_I}$ is the line profile

$$r_{\mathcal{F}_I} = 1 - (\mathcal{F}_I / \mathcal{F}_{I_c}). \tag{2.38}$$

$\mathcal{F}_I$ (resp. $\mathcal{F}_{I_c}$) is the integral over the visible stellar disk of the emergent intensity in the line (resp. in the neighbouring continuum). Importantly, contrary to the mean longitudinal magnetic field, the mean quadratic magnetic field provides a measurement of the field strength that is fairly insensitive to its structure. Thus it is especially well suited to detect fields that have a complex structure.

The crossover effect can be measured by the second-order moment about the centre of the profiles of spectral lines recorded in Stokes V. Mathys (1995) showed that one can derive from the measurements a quantity called the mean asymmetry of the longitudinal magnetic field, which is the first moment of the component of the magnetic field along the line of sight, about the plane defined by the line of sight and the stellar rotation axis. The second-order moment of a line profile recorded in Stokes V with respect to the wavelength λ_0 of the centre of gravity of the corresponding unpolarised profile is defined as

$$R_V^{(2)}(\lambda_0) = \frac{1}{W_\lambda} \int r_{\mathcal{F}_V}(\lambda - \lambda_0)(\lambda - \lambda_0)^2 d\lambda, \tag{2.39}$$

where W_λ is the equivalent width of the line and $r_{\mathcal{F}_V}$ is the line profile in Stokes V.

An application of the moment technique for magnetic field studies can be found in Figure 2.6, which shows the results for the Ap star HD 137909 (Mathys & Hubrig 1997).

2.9 Low-resolution Correlations

Spectropolarimetry, and in particular observations of spectral lines in right and left circular polarisation, has for more than 70 years been a privileged tool to diagnose mean longitudinal magnetic fields. Currently, observing the spectrum of a star in the two senses of circular polarisation and forming the V/I difference spectrum is in almost all cases the most sensitive method of detecting a weak magnetic field on the surface of a star. The longitudinal magnetic field is especially sensitive against the aspect of the magnetic field seen, as explained in the framework of the oblique rotator model (see Section 2.3).

One method to determine the mean longitudinal magnetic field $\langle B_z \rangle$ was originally proposed by Angel & Landstreet (1970) and adapted for the low-resolution spectropolarimeter FORS 1 (now FORS 2; Appenzeller et al. 1998) by Bagnulo et al. (2001). The mean longitudinal magnetic field $\langle B_z \rangle$ is diagnosed from the slope of a linear regression.

FORS 2 is a multimode instrument equipped with polarisation-analysing optics comprising super-achromatic half-wave and quarter-wave phase retarder plates, and a Wollaston prism with a beam divergence of 22″ in standard-resolution mode. To measure magnetic fields, left- and right-hand circular polarisation data is obtained in the ordinary and the extraordinary beams, two spectra on the same detector. The position angle of the quarter-wave retarder plate is alternated in the sequence $-45°$ $+45°+45°-45°$, swapping the left- and right-hand circular polarisation beams on the detector. Typically, between two and six such sequences of exposures are combined to form a single polarimetric observation. One of the most used dispersing elements is the GRISM 600B, which together with the narrowest available slit width of 0.4″ leads to a spectral resolving power of $R \approx 2000$. FORS 2 provides, in combination with one of the 8 m diameter mirrors of the Very Large Telescope, the high

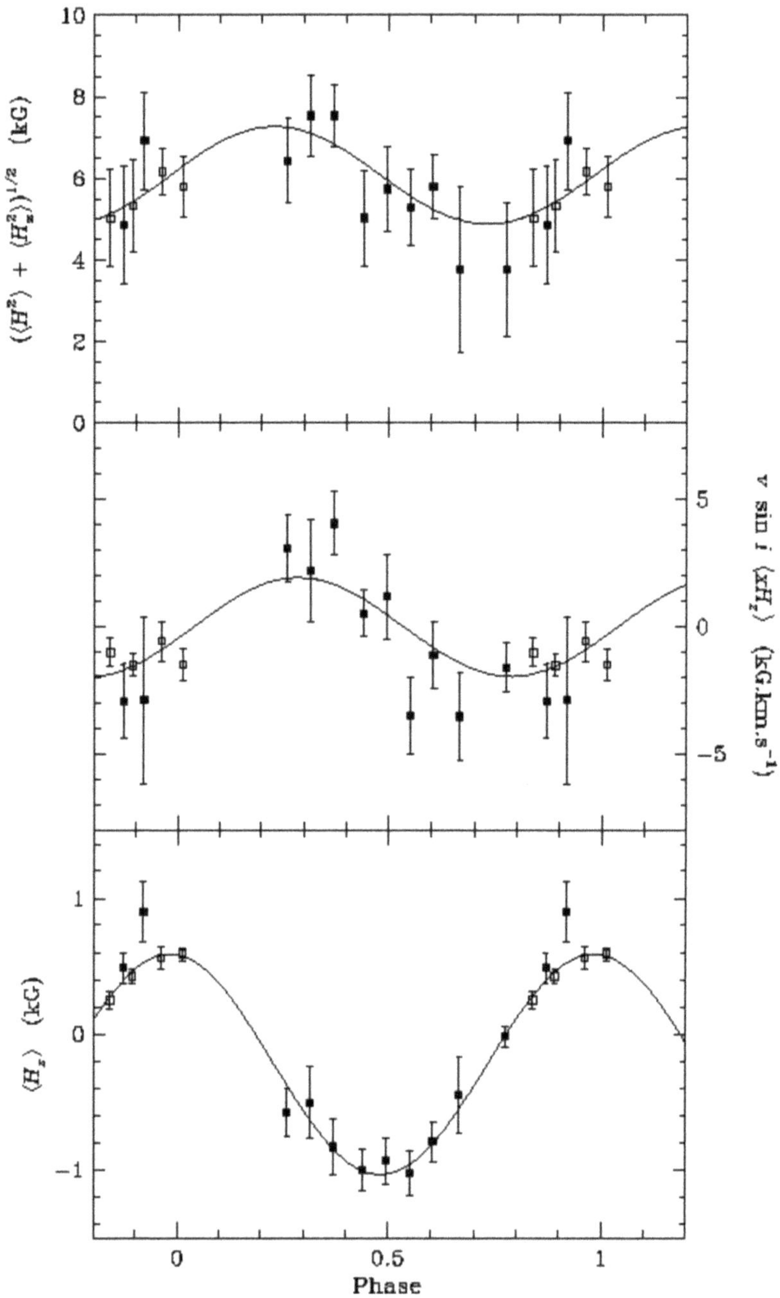

Figure 2.6. Phase diagram of measurements of the mean longitudinal magnetic field (*lower panel*), of the crossover (*middle panel*) and of the mean quadratic magnetic field (*upper panel*) of HD 137909 (the symbols are explained in Mathys & Hubrig 1997). The curves are least-squares fits of the data by sinusoids. Please note that the authors use *H* in the same way *B* is used in this book (see Equations (2.35) and (2.36)). Credit: Mathys & Hubrig (1997), reproduced with permission © ESO.

sensitivity needed to determine magnetic fields in faint targets, using rather short exposure times.

The determination of the mean longitudinal magnetic field using low-resolution FORS 1/2 spectropolarimetry has been described in detail in papers by Bagnulo et al. (2002, 2009, 2012) and Hubrig et al. (2004a, 2004b, 2016).

The mean longitudinal magnetic field is determined through a regression analysis between the Stokes V and Stokes I spectra, following:

$$\frac{V}{I} = -\frac{g_{\mathrm{eff}}\, e\, \lambda^2}{4\pi\, m_{\mathrm{e}}\, c^2}\, \frac{1}{I}\, \frac{\mathrm{d}I}{\mathrm{d}\lambda} \langle B_z \rangle, \tag{2.40}$$

where V is the Stokes parameter that measures the circular polarisation, I is the intensity in the unpolarised spectrum, g_{eff} is the effective Landé factor, e is the electron charge, λ is the wavelength, m_{e} is the electron mass, c is the speed of light, $\mathrm{d}I/\mathrm{d}\lambda$ is the wavelength derivative of Stokes I, and $\langle B_z \rangle$ is the mean longitudinal magnetic field.

After the extraction of the ordinary and extraordinary spectrum for each individual raw frame, the V/I spectrum is calculated using

$$\frac{V}{I} = \frac{1}{2}\left\{\left(\frac{f^{\mathrm{o}} - f^{\mathrm{e}}}{f^{\mathrm{o}} + f^{\mathrm{e}}}\right)_{-45^\circ} - \left(\frac{f^{\mathrm{o}} - f^{\mathrm{e}}}{f^{\mathrm{o}} + f^{\mathrm{e}}}\right)_{+45^\circ}\right\}, \tag{2.41}$$

and the I spectrum following

$$I = \frac{1}{4}\{(f^{\mathrm{o}} + f^{\mathrm{e}})_{-45^\circ} + (f^{\mathrm{o}} + f^{\mathrm{e}})_{+45^\circ}\}, \tag{2.42}$$

where $+45^\circ$ and -45° indicate the position angle of the quarter-waveplate and f^{o} and f^{e} are the fluxes of the ordinary and extraordinary beams, respectively. The double difference in Equation (2.41) ensures the removal of instrumental polarisation.

Additionally, the diagnostic N parameter is calculated following the formalism of Bagnulo et al. (2009):

$$N_{m,n} = \left(\frac{V}{I}\right)_m - \left(\frac{V}{I}\right)_n \tag{2.43}$$

with $N_{m,n}$ the diagnostic null parameter determined from the mth and nth pair of V/I spectra, $(V/I)_m$ and $(V/I)_n$, respectively. The diagnostic null parameter N is combined in such a way that the real polarisation signal is expected to cancel out.

Before performing the regression analysis, a few steps should be performed to remove undesired effects in the data and to make an independent error estimate. The null profiles N are calculated as pairwise differences from all available V/I profiles. From the diagnostic N profiles, 3σ-outliers are identified and used to clip the Stokes V profiles. This removes spurious signals, which mostly come from cosmic rays, and also reduces the noise. A rectification of the Stokes V/I spectra is needed in FORS 1/2 observations of different types of objects such as Herbig Ae/Be stars with a complex circumstellar environment, WR stars with strong dense winds, O-type stars with magnetospheres, etc. The first detection of the presence of variable slopes in the continua of the *V*/*I* spectra in FORS 1 observations of the X-ray binary Cyg X-1 was reported by Karitskaya et al. (2009, 2010). For the rectification, fitting and subtracting a line or a low-order polynomial to the data is performed. Starting in 2010, Monte-Carlo bootstrapping tests were implemented in the analysis of FORS 2 measurements (Rivinius et al. 2010; Steffen et al. 2014; Hubrig et al. 2014). These are most often applied with the purpose of deriving robust estimates of standard errors.

An example of the application of this method is illustrated in Figure 2.7 for the B2IV star CPD −62 2124 with $m_V = 10.99$, which led to $\langle B_z \rangle = 5222 \pm 123$ G from the hydrogen lines (see Schöller et al. 2017).

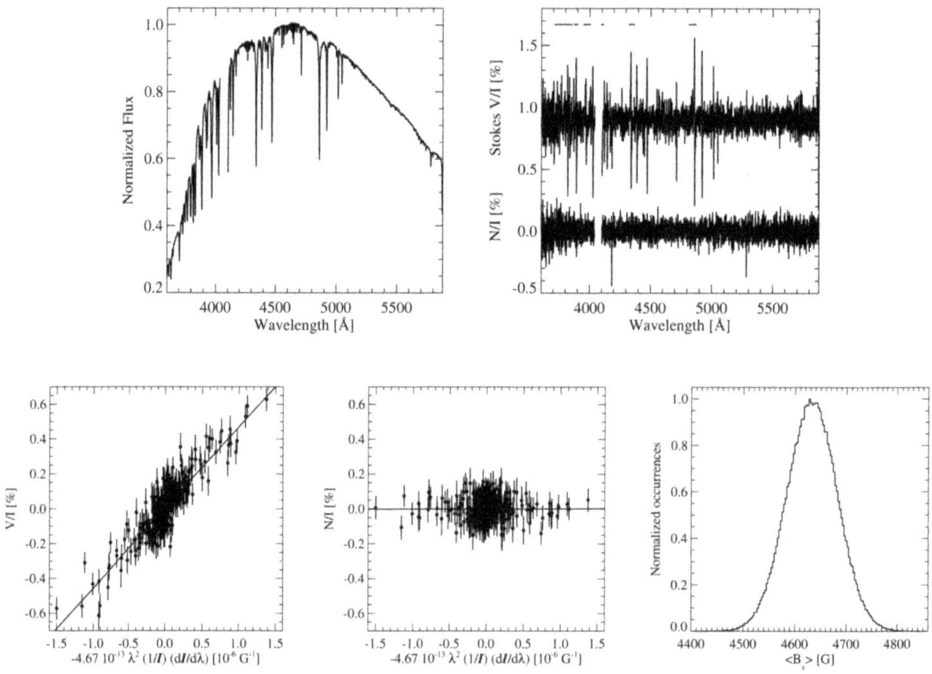

Figure 2.7. The analysis of FORS 2 data of CPD −62 2124, collected during the night of 2015 March 17, considering the hydrogen lines. *Top left panel*: observed Stokes *I* spectrum arbitrarily normalised to the highest value. *Top right panel*: the top profile shows Stokes *V* (in %), while the bottom profile shows the spectrum of the *N* parameter (in %). The Stokes *V* spectrum is shifted upward by 0.9% for better visibility. The regions used to calculate the longitudinal magnetic field are indicated by horizontal lines close to the top of the panel. *Bottom left panel:* linear fit to Stokes *V*. *Bottom middle panel:* linear fit to the *N* spectrum. From the linear fit, one can determine $\langle N_z \rangle = 28 \pm 102$ G. *Bottom right panel:* distribution of the longitudinal magnetic field values $P(\langle B_z \rangle)$, which were obtained via bootstrapping. From the distribution $P(\langle B_z \rangle)$, one obtains the most likely value for the longitudinal magnetic field $\langle B_z \rangle = 4636 \pm 128$ G. The gaps in the region around Hδ in the two upper panels result from masking an internal reflection in that spectral range. Credit: Schöller et al. (2017), reproduced with permission © ESO.

2.10 LSD and SVD Techniques

Zeeman signatures in the spectra of many stars are very small and it is not always easy to detect and interpret them. Increasing the signal-to-noise ratio (S/N) by increasing the exposure time is frequently limited by short rotation periods or pulsations, which both smear out the signal. Thus, multi-line approaches, as first proposed by Semel (1989), are commonly used to increase the S/N.

The most widely used approach is the least squares deconvolution (LSD; Donati et al. 1997). Under the assumption of the weak-field approximation, i.e., the magnetic splitting of spectral lines is assumed to be smaller than their Doppler broadening, line-profile shapes are considered self-similar and amplitudes only linearly depend upon the individual Landé factors and line strengths. Thus, the entire spectrum can be described as a sum of scaled and shifted identical profiles, which can be deconvolved to create an average profile in velocity space. Due to non-linear effects in the summation and the effect of blends, the resulting LSD profiles are not observed, but rather processed Zeeman signatures. The drawback of this method is that any intrinsic line-to-line difference is wiped out in the process. In particular, all transitions are treated as if they had the same Zeeman pattern. The limited validity of this assumption when lines with very different Zeeman patterns are analysed has an unknown impact on the derived values of the mean longitudinal magnetic field. Using this method, special care should be taken in studies of hot massive stars or pre-main sequence stars due to the rather low number of spectral lines in the optical spectral region, which are, in addition, to a different extent subjected to wind or accretion signatures. Kochukhov et al. (2010) indicated that the Stokes LSD profiles do not behave as a real spectral line with respect to the magnetic field and the element abundance. They concluded that the usual method of interpreting the LSD profiles, assuming that they are equivalent to a real spectral line, gives satisfactory results only in a limited parameter range, and presented an improved version of LSD, which they called iLSD.

Sennhauser & Berdyugina (2010) presented a method they called Zeeman component decomposition (ZCD) for recovering an intrinsic line profile and the mean longitudinal magnetic field from polarised spectra. ZCD can be deployed with little prior information on the stellar parameters by fitting each atomic/molecular line depth independently and thus take out some of the profile nonlinearities due to blending. A comparison of their first results with results from Grunhut et al. (2010), which were based on LSD, showed that ZCD magnetic-field strengths were generally a factor of two smaller than from LSD. This was attributed by Sennhauser & Berdyugina (2010) to LSD being more prone to line blending than ZCD.

Another method for Stokes profiles extraction was proposed by Carroll et al. (2007) and Martínez González et al. (2008), using the principal component analysis (PCA). A much simpler, but also very effective coherent addition of line profiles in the velocity or logarithmic wavelength domain was put forward by Semel et al. (2009) and Ramírez Vélez et al. (2010). A new two-step Ansatz was described by Carroll et al. (2012) in their application to the T Tauri star V410 Tau. It is called the subspace reconstruction method and based on a singular value decomposition (SVD) of the observation matrix. From a large set of suitable and Zeeman-sensitive spectral line profiles (so far only Stokes V), an observation matrix in the velocity domain is built. To search for systematic and correlated features common to all Stokes V profiles, an SVD analysis is applied to the observation matrix. The decomposition into eigenprofiles of the observation covariance matrix provides an efficient method for separating coherent and systematic structures present in all line profiles from non-coherent random features, e.g., noise and blends. The number of significant eigenprofiles determines the signal subspace (the reduced rank of the observation covariance matrix) that carries the majority of the profile information. In a second stage, which provides the actual signal boosting effect, a signal averaging in the subspace is performed. This is done by projecting all Stokes V line profiles onto the limited number of eigenprofiles (dimensions) and eventually averaging over the projection coefficients. The effect of this two-stage process is—besides the actual signal extraction—a better noise suppression compared to conventional line profile averaging or LSD. This method has the great advantage that it facilitates a detailed radiative transfer based line-profile fitting. As an example, Figure 2.8 shows SVD profiles of the Herbig Ae star HD 95881 possessing a weak magnetic field of the order of 100 G (from Järvinen et al. 2018).

The mean longitudinal magnetic field is determined by computing the first-order moment of the Stokes V profile determined by one of the techniques described above according to Mathys (1989):

$$\langle B_z \rangle = -2.14 \times 10^{11} \frac{\int vV(v)dv}{\lambda_0 g_0 c \int [I_c - I(v)]dv}, \tag{2.44}$$

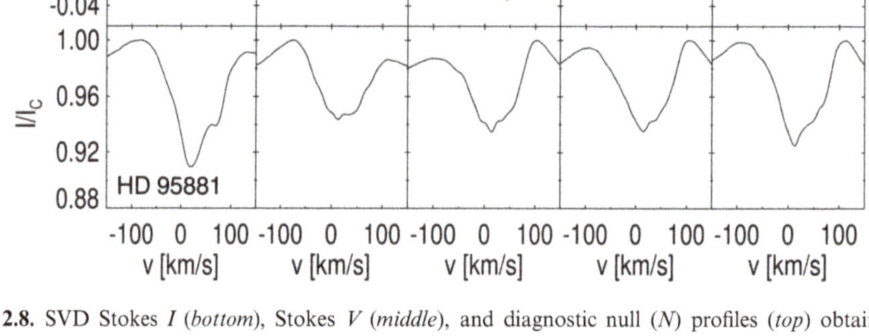

Figure 2.8. SVD Stokes I (*bottom*), Stokes V (*middle*), and diagnostic null (N) profiles (*top*) obtained for HD 95881. The Stokes V and N profiles have been amplified by a factor of 50. The profiles are plotted corresponding to the increasing date. The horizontal dashed lines indicate the average values and the $\pm 1\sigma$-ranges. The detected Zeeman features are highlighted in red. Credit: Figure 2 from Järvinen et al. (2018).

where v is the velocity shift from the line centre, in km s^{-1}, and λ_0 and g_0 are the average values of, respectively, the wavelengths (in nm) and the effective Landé factors of all the lines used to compute the LSD profile.

The classification of the quality of the magnetic field detections makes use of the false alarm probability (FAP; Donati et al. 1992), rather than the measurement error. Donati et al. (1992) considered a profile with FAP $< 10^{-5}$ as a definite detection, with $10^{-5} <$ FAP $< 10^{-3}$ as a marginal detection, and FAP $> 10^{-3}$ as a non-detection.

2.11 Zeeman–Doppler Imaging

Zeeman–Doppler imaging (ZDI) techniques extract information on both the stellar chemical structure and the magnetic topology from observations of the rotational modulation of line profiles observed in polarised light. ZDI converts a set of line profiles observed over the stellar rotation period, where each line profile represents an instantaneous one-dimensional projection of the stellar surface, into a two-dimensional map. This inversion technique, which can be applied to spectral time-series obtained for two or four Stokes parameters, was introduced by Semel (1989). Depending on how the map is parameterised, it requires an additional mathematical constraint—a regularisation—to ensure the uniqueness of the solution. ZDI performs best for fast rotating stars with substantial Doppler broadening.

As the employment of individual line profiles requires spectropolarimetric observations of very high quality, the average Stokes parameter profiles are usually calculated with LSD or SVD. The Zeeman–Doppler imaging code iMap written by Carroll et al. (2007, 2009a) permits to simultaneously reproduce the temperature and magnetic vector field distribution from a sequence of observed Stokes I and V profiles. One of the key features of the line profile reconstruction is that the SVD line profiles are amenable to radiative transfer modelling within the iMap code (Carroll et al. 2007, 2012). The code also utilises a new iterative regularisation scheme, which is independent of any additional surface constraints (Carroll et al. 2007, 2012). The ZDI map for the T Tauri star V410 Tau is presented as an orthographic plot in Figure 2.9. A potential field source surface extrapolation for this star calculated from the Zeeman–Doppler map is shown in Figure 2.10.

So far, ZDI was applied for only a few Ap/Bp stars (Kochukhov 2020). The global magnetic field geometry changes little from one star to another, with nearly

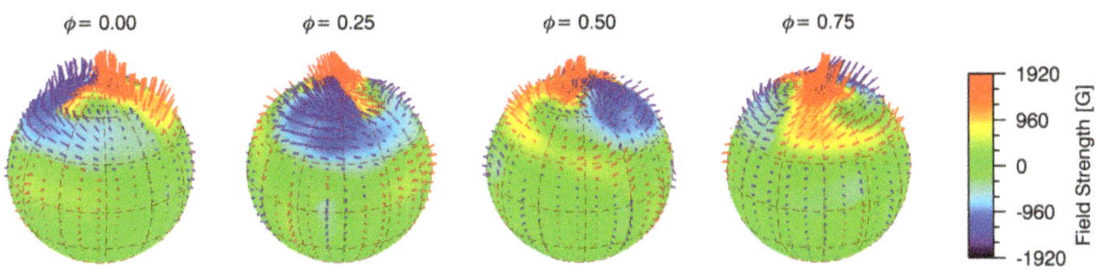

Figure 2.9. Orthographic maps of the magnetic field distribution of V410 Tau at four different rotational phases ϕ. The magnetic field is colour coded and the field lines are proportional to the field strength (blue negative and red positive polarity). Credit: Carroll et al. (2012), reproduced with permission © ESO.

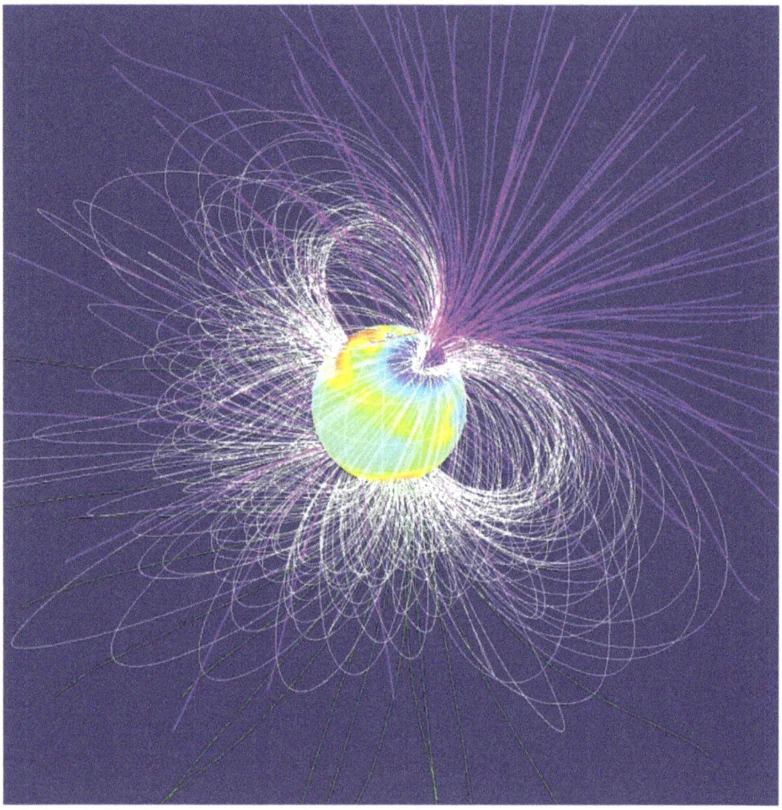

Figure 2.10. A potential field source surface extrapolation for the weak-line T Tauri star V410 Tau. The extrapolation shows the open and closed field lines calculated from the Zeeman–Doppler map of V410 Tau. Credit: Figure courtesy of Thorsten A. Carroll.

all stars showing dominant dipolar fields with a varying degree of distortion. The application of the ZDI method requires an in-depth understanding of the stellar atmospheres of magnetic stars. Due to the 3D nature of the atomic diffusion process in magnetic Ap/Bp star atmospheres and an inadequate choice of the regularisation function, the detectability of a variety of abundance structures by means of ZDI is certainly limited and can lead to multiple solutions for abundance maps (e.g., Alecian & Stift 2019). This was illustrated when the same polarimetric spectra for the same target were analysed by two different groups with different software packages using magnetic Doppler imaging (Kochukhov et al. 2009) and iMap (Carroll et al. 2009b). Their maps showed completely different surface magnetic field topologies, while the distribution of temperature spots appeared similar.

References

Alecian, G., & Stift, M. J. 2019, MNRAS, 482, 4519

Angel, J. R. P., & Landstreet, J. D. 1970, ApJL, 160, L147

Appenzeller, I., Fricke, K., Fürtig, W., et al. 1998, Msngr, 94, 1

Bagnulo, S., Szeifert, T., Wade, G. A., Landstreet, J. D., & Mathys, G. 2001, Msngr, 104, 32

Bagnulo, S., Szeifert, T., Wade, G. A., Landstreet, J. D., & Mathys, G. 2002, A&A, 389, 191

Bagnulo, S., Landolfi, M., Landstreet, J. D., et al. 2009, PASP, 121, 993

Bagnulo, S., Landstreet, J. D., Fossati, L., & Kochukhov, O. 2012, A&A, 538, A129

Beckers, J. M. 1969, A Table of Zeeman Multiplets, Sacramento Peak Observatory Contribution no.141, Physical Science Research Papers no. 371 (Sacramento, CA: Peak Observatory and Air Force Cambridge Research Laboratories)

Borra, E. F., & Deschatelets, D. 2015, AJ, 150, 146

Carroll, T. A., Kopf, M., Ilyin, I., & Strassmeier, K. G. 2007, AN, 328, 1043

Carroll, T. A., Kopf, M., Strassmeier, K. G., & Ilyin, I. 2009a, in IAU Symp. 259, Cosmic Magnetic Fields: From Planets, to Stars and Galaxies, ed. K. G. Strassmeier, A. G. Kosovichev, & J. E. Beckman (Cambridge: Cambridge Univ. Press), 633

Carroll, T. A., Kopf, M., Strassmeier, K. G., Ilyin, I., & Tuominen, I. 2009b, in IAU Symp. 259, Cosmic Magnetic Fields: From Planets, to Stars and Galaxies, ed. K. G. Strassmeier, A. G. Kosovichev, & J. E. Beckman (Cambridge: Cambridge Univ. Press), 437

Carroll, T. A., Strassmeier, K. G., Rice, J. B., & Künstler, A. 2012, A&A, 548, A95

Donati, J. F., Semel, M., & Rees, D. E. 1992, A&A, 265, 669

Donati, J. F., Semel, M., Carter, B. D., Rees, D. E., & Collier Cameron, A. 1997, MNRAS, 291, 658

Grunhut, J. H., Wade, G. A., Hanes, D. A., & Alecian, E. 2010, MNRAS, 408, 2290

Hanle, W. 1923, NW, 690

Hubrig, S., Cowley, C. R., Bagnulo, S., et al. 2002, in ASP Conf. Ser. 279, Exotic Stars as Challenges to Evolution ed. C. A. Tout, & W. van Hamme (San Francisco, CA: ASP), 365

Hubrig, S., Kurtz, D. W., Bagnulo, S., et al. 2004a, A&A, 415, 661

Hubrig, S., Szeifert, T., Schöller, M., Mathys, G., & Kurtz, D. W. 2004b, A&A, 415, 685

Hubrig, S., Schöller, M., & Kholtygin, A. F. 2014, MNRAS, 440, 1779

Hubrig, S., Scholz, K., Hamann, W. R., et al. 2016, MNRAS, 458, 3381

Ignace, R., Nordsieck, K. H., & Cassinelli, J. P. 1997, ApJ, 486, 550

Järvinen, S. P., Carroll, T. A., Hubrig, S., et al. 2018, ApJL, 858, L18

Järvinen, S. P., Hubrig, S., Schöller, M., et al. 2021, MNRAS, 501, 4534

Jones, R. C. 1941, JOSA, 31, 500

Karitskaya, E. A., Bochkarev, N. G., Hubrig, S., et al. 2009, arXiv:0908.2719

Karitskaya, E. A., Bochkarev, N. G., Hubrig, S., et al. 2010, IBVS, 5950, 1

Kochukhov, O., Piskunov, N., Ilyin, I., & Tuominen, I. 2009, in IAU Symp. 259, Cosmic Magnetic Fields: From Planets, to Stars and Galaxies, ed. K. G. Strassmeier, A. G. Kosovichev, & J. E. Beckman (Cambridge: Cambridge Univ. Press), 439

Kochukhov, O., Makaganiuk, V., & Piskunov, N. 2010, A&A, 524, A5

Kochukhov, O. 2020, in Proc. Polish Astronomical Soc. 11, Stellar Magnetism: A Workshop in Honour of the Career and Contributions of John D. Landstreet, ed. G. Wade, E. Alecian, D. Bohlender, & A. Sigut, 89

Landolfi, M., Landi Degl'Innocenti, E., Landi Degl'Innocenti, M., & Leroy, J. L. 1993, A&A, 272, 285

Landolfi, M., Bagnulo, S., Landi Degl'Innocenti, M., & Landi Degl'Innocenti, E. 2001, in ASP Conf. Ser. 248, Magnetic Fields Across the Hertzsprung–Russell Diagram ed. G. Mathys, S. K. Solanki, & D. T. Wickramasinghe (San Francisco, CA: ASP), 349

Leroy, J. L. 1989, A&A, 215, 360

Müller, H. 1943, Memorandum on the Polarization Optics of the Photo-elastic Shutter, Technical Report, Report Number 2 of the OSRD Project OEMsr-576
Martínez González, M. J., Asensio Ramos, A., Carroll, T. A., et al. 2008, A&A, 486, 637
Mathys, G., & Hubrig, S. 1997, A&AS, 124, 475
Mathys, G., & Hubrig, S. 2006, A&A, 453, 699
Mathys, G., Hubrig, S., Landstreet, J. D., Lanz, T., & Manfroid, J. 1997, A&AS, 123, 353
Mathys, G. 1988, A&A, 189, 179
Mathys, G. 1989, FCPh, 13, 143
Mathys, G. 1990, A&A, 232, 151
Mathys, G. 1993, in ASP Conf. Ser. 44, IAU Colloq. 138: Peculiar versus Normal Phenomena in A-type and Related Stars ed. M. M. Dworetsky, F. Castelli, & R. Faraggiana (San Francisco, CA: ASP), 232
Mathys, G. 1995, A&A, 293, 733
Mathys, G. 2017, A&A, 601, A14
Preston, G. W. 1967, ApJ, 150, 547
Preston, G. W. 1969, ApJ, 156, 967
Preston, G. W. 1971, ApJ, 164, 309
Ramírez Vélez, J. C., Semel, M., Stift, M., et al. 2010, A&A, 512, A6
Rivinius, T., Szeifert, T., Barrera, L., et al. 2010, MNRAS, 405, L46
Schöller, M., Hubrig, S., Fossati, L., et al. 2017, A&A, 599, A66
Semel, M., Ramírez Vélez, J. C., Martínez González, M. J., et al. 2009, A&A, 504, 1003
Semel, M. 1989, A&A, 225, 456
Sennhauser, C., & Berdyugina, S. V. 2010, A&A, 522, A57
Steffen, M., Hubrig, S., Todt, H., et al. 2014, A&A, 570, A88
Stenflo, J. O. 1982, SoPh, 80, 209
Stibbs, D. W. N. 1950, MNRAS, 110, 395
Stokes, G. G. 1851, TCaPS, 9, 399
Zeeman, P. 1897, ApJ, 5, 332

Swetlana Hubrig and Markus Schöller

Chapter 3

Observing Magnetic Fields

Magnetic fields are difficult to detect since they can rarely be observed directly. To probe magnetic fields in different types of stars, astronomers usually rely on observations of polarised light using diverse observational techniques and analysis tools. The Zeeman components of spectral lines are polarised and so permit stellar magnetic fields to be measured. For a long time, magnetic Ap/Bp stars were the only non-degenerate stars besides the Sun in which direct detections of magnetic fields had been achieved. These stars generally possess large-scale organised magnetic fields, most frequently dominated by a dipole, that are inferred from the circular polarisation in spectral lines. Only during the last decade has observational evidence been accumulating that other non-degenerate stars with radiative envelopes can harbour magnetic fields. Currently, a particular shortcoming in the studies of magnetic fields is the paucity of analysing equipment for polarised light at high and low-resolution spectrographs installed at telescopes with large apertures. Thus, the best knowledge of the magnetic field strength and geometry is limited to bright Ap/Bp stars.

3.1 Stellar Characteristics as Indirect Magnetic Field Tracers

A considerable fraction of upper main sequence A and B stars, typically 10%–20% (e.g., 15%; Preston 1974), exhibits atmospheric chemical anomalies that, apparently, do not have an evolutionary origin, but rather reflect the operation of various competing physical effects, such as e.g., microscopic diffusion of trace atomic species, which may act in interaction with a magnetic field. It is rather easy to recognise these stars already in low-resolution spectra because their distinctive atmospheric chemical anomalies are a function of effective temperature $T_{\rm eff}$. The coolest Ap stars with the lowest mass usually show overabundances of Sr, Cr, and rare earths, and higher mass late-Bp stars have deficient He and overabundant Si. The most massive chemically peculiar Bp stars show overabundant He. On stellar surfaces of these stars, which have large-scale magnetic fields, different chemical

doi:10.1088/2514-3433/abefccch3

elements are usually inhomogeneously distributed in a way apparently related to the geometry of their magnetic fields. While rare-earth elements seem to concentrate toward the magnetic poles, other chemical elements, such as Si, Ca, Cr, and Fe are distributed more uniformly over the stellar surfaces (e.g., Nesvacil et al. 2012). Notably photometric, spectroscopic, and magnetic variability in magnetic stars present an unambiguous diagnostic of their rotation periods. The spots in Ap/Bp stars are known to be stable over decades, hence it is possible to see rotational variability caused by the varying aspect of the spots as long as the geometry of the observation is such that substantially different regions of the stellar surface come into view at different rotation phases.

Ap/Bp stars rotate more slowly than their normal counterparts, sometimes by five to six orders of magnitude (e.g., Mathys 2017). From consideration of the distribution of the periods of Ap stars with spectral lines resolved into their magnetically split components, Mathys (2017) argued that several percent of the Ap stars must have rotation periods longer than one year and that some of them definitely have periods of the order of 300 yr. Even Ap stars with much longer periods, up to ~1000yr, may exist. The fastest rotating Ap stars have periods of the order of half a day (e.g., Mathys 2004; Netopil et al. 2017).

Magnetic cool Ap stars and Ap stars showing rapid oscillation (Kurtz 1982) can be easily recognised on medium and high-resolution spectra as they show a strong core-wing anomaly in the hydrogen lines, particularly the $H\alpha$ and $H\beta$ lines (Cowley et al. 2001). Broad wings of these lines, indicative of temperatures in the range 7000–8000 K end abruptly in narrow cores. The widths of the line cores are compatible with those of dwarfs with temperatures of 6000 K or lower. This profile appearance has been known for a long time for Przybylski's star, which has a large excess of almost all rare-earth elements (Przybylski 1961), but it is seen in other cool Ap stars and indicates an abnormal atmospheric structure. Strong deviations of Ap stellar atmospheric models from normal A star models are also manifested in the ionisation imbalance of rare-earth elements. It was shown by, e.g., Ryabchikova et al. (2002) that no consistent abundances for the second and third ionisation states of these elements can be derived for roAp stars, suggesting a strong vertical stratification of these elements. Significant vertical stratification effects for many other elements were found in hotter magnetic Ap/Bp stars, but also in non-magnetic or weakly magnetic Bp stars (e.g., Bohlender 2005; Castelli & Hubrig 2007; Catanzaro et al. 2016; Sadakane & Nishimura 2018).

Many unexplained phenomena, such as cyclic behaviour on a rotational timescale observed in the UV wind lines (e.g., Henrichs et al. 2000), $H\alpha$ emission variation (e.g., Donati et al. 2002), chemical peculiarity (e.g., Reiners et al. 2000), the presence of narrow X-ray emission lines (e.g., Cohen et al. 2003) or non-thermal radio/X-ray emission (e.g., Bieging et al. 1989), were observed in massive stars and were considered as an indirect observational evidence for the presence of magnetic fields. Additionally, a number of OB-type stars showed nitrogen enrichment (e.g., Morel et al. 2008; Hubrig et al. 2008), indicating that magnetic fields are probably present in these stars. As an example, one of the hottest β Cephei stars, the B0.7IV star ξ^1 CMa has been studied by Morel et al. (2006), who derived abundances of several elements including nitrogen, and found that ξ^1 CMa is nitrogen enriched. FORS 1 spectropolarimetric observations of this star indeed showed the presence of a rather strong longitudinal magnetic field of up to 400 G (Hubrig et al. 2006a). FORS 1 Stokes V spectra of ξ^1 CMa around the He II line at 4121 Å and the Hβ line are presented in Figure 3.1. However, subsequent magnetic studies showed that not all nitrogen-rich stars possess detectable magnetic fields (Morel et al. 2008; Hubrig et al. 2008).

The role of X-ray emission as an indirect indicator for the presence of magnetic fields in massive stars is not clear yet. Anomalous X-ray emission can be expected from a massive star with a magnetic field strong enough to regulate the stellar wind motion. The relative importance of magnetic fields in gases is described by the plasma-β_p parameter with $\beta_p = 8\pi p/B^2$, where p is the gas pressure. The gas is magnetically dominated when $\beta_p < 1$. For supersonic flows, such as stellar winds, the ram pressure exceeds the gas pressure and the dynamical importance of a magnetic field is defined by the ratio of wind kinetic to magnetic energy density: when this ratio is less than one, the magnetic field dominates the wind bulk motion. While some magnetic massive stars show bright, hard, variable X-ray emission as predicted by the magnetically confined wind-shock scenario (e.g., Schulz et al. 2000), others are relatively soft X-ray sources (Nazé et al. 2010). Variations of the longitudinal magnetic field, of broad-band photometry, of visible emission lines and of the X-ray flux occur simultaneously in a few magnetic O-type stars (e.g., Donati et al. 2002, 2006; Nazé et al. 2007). For some other magnetic massive stars, the X-ray emission is very faint, or no significant short-term or long-term variation of the X-ray flux could be detected (e.g., Nazé et al. 2008). In some stars showing strong X-ray emission, no magnetic field detections were achieved (e.g., Nazé et al. 2012). Thus, while strong hard X-ray emission serves as a good indicator for the potential presence of a strong stellar magnetic field, X-ray strong stars represent only a sub-class of all magnetic massive stars.

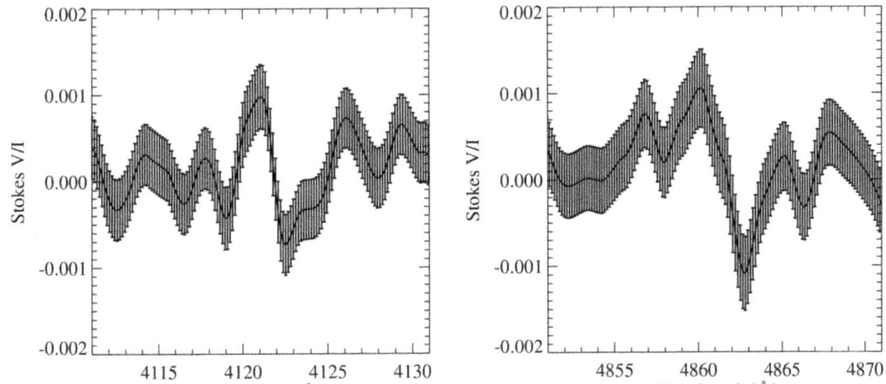

Figure 3.1. Stokes V/I spectra of the β Cephei star ξ^1 CMa in the vicinity of the He II line at 4121 Å (*left panel*) and the Hβ line (*right panel*). The thickness of the plotted line corresponds to the uncertainty of the polarisation measurement determined from photon noise. Credit: Figure 1 and 2 from Hubrig (2008).

One of the most obvious characteristics of magnetic massive early B- and O-type stars is the presence of strong spectral variability, which can already be detected in low-resolution spectra. While a major characteristic of Ap/Bp stars is the spectral variability due to the presence of chemical spots showing a certain symmetry with respect to the magnetic field geometry, the interpretation of line profile variations in massive stars seems not to be as simple: the behaviour of line profiles of different ions of a given element does not clearly correspond to abundance variations due to chemical spots. Synthetic spectra calculations of line profile variations in massive stars are more troublesome since different spectral lines are to a different degree sensitive to the stellar wind and to non-LTE effects. A few examples of long- and short-term variability of spectral line profiles in low-resolution ($R \approx 2000$) FORS 2 spectra of the magnetic Of?p star CPD−28 2561 are presented in Figure 3.2 (Hubrig et al. 2013). The magnetic field of the order of a few hundred Gauss in this star was detected by Hubrig et al. (2011b). Apart from the He I λ4026 line, which showed only a weak variability of about 1%–2% of the continuum level, three other lines, He I λ5876, He II λ4686, and Hβ, showed a strongly variable contribution of an emission component. The line He II λ4686 exhibited strong emission during observations in 2010, which transformed to a much weaker P Cygni line profile with a weak blueshifted absorption in 2011. A contribution of the emission component is clearly seen in the profile of the He I λ5876 line in 2010, but its presence could only be suspected in the variable line depth observed in 2011. A strong emission component

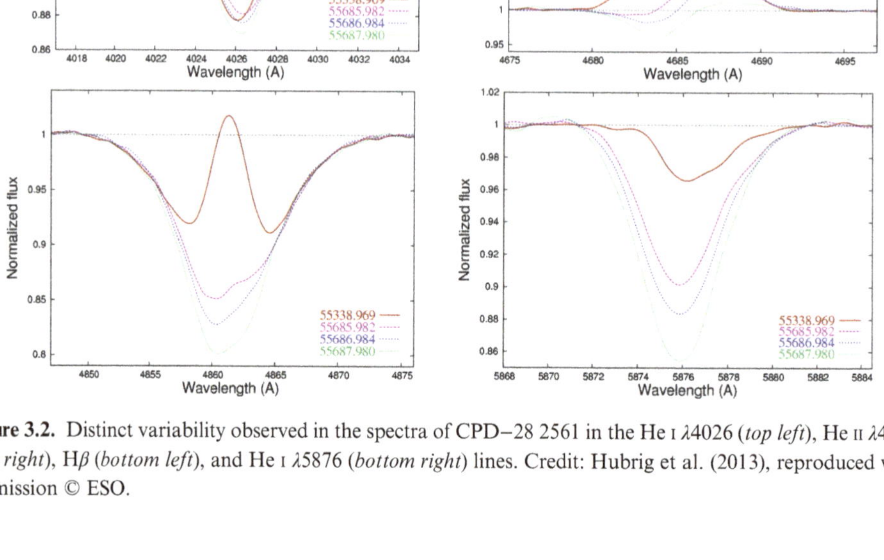

Figure 3.2. Distinct variability observed in the spectra of CPD−28 2561 in the He I λ4026 (*top left*), He II λ4686 (*top right*), Hβ (*bottom left*), and He I λ5876 (*bottom right*) lines. Credit: Hubrig et al. (2013), reproduced with permission © ESO.

was well visible in the line centre of Hβ in 2010. CPD−28 2561 is an intrinsically bright X-ray source as detected by the ROSAT X-ray observatory. The ratio of the X-ray luminosity to the bolometric luminosity ($\log L_X/L_{bol} \approx -6$) makes it unusually X-ray bright compared to other O-type stars (e.g., Chlebowski 1989). The hardness ratio (HR1 $\approx 0.04 \pm 0.4$) indicates that its X-ray emission can be harder than is typical in single O-type stars.

Also a detection of ejection nebulae associated with a few massive stars can probably be considered as an indirect tracer of the presence of a magnetic field. The magnetic Of?p star HD 148937 is surrounded by the bipolar nebula NGC 6164/5, which displays an axisymmetrical geometry. Due to its anomalous chemical abundances, this nebula is thought to have formed during an LBV-like outburst and expands with a projected velocity of about 30 km s^{-1} (Leitherer & Chavarria-K 1987). The nebula must have been ejected less than 2×10^5 yr ago and shows a nitrogen overabundance of a factor five compared to that of the Sun and the Orion nebula. The most recent study of the NGC 6164/5 nebula by Mahy et al. (2017) using infrared Spitzer IRS and far-infrared Herschel/PACS observations revealed that its structure is particularly complex and is composed, from the centre on outwards, of a close bipolar ejecta nebula (NGC 6164/5), an ellipsoidal wind-blown shell and a spherically symmetric Strömgren sphere. The combined analyses of the known kinematics and of the abundances of the nebula suggested either a helical morphology for the nebula, possibly linked to the magnetic geometry, or the occurrence of a binary merger. The merger scenario for the generation of magnetic fields in upper main sequence stars was previously suggested by Tout et al. (2008), Ferrario et al. (2009), and Tutukov & Fedorova (2010).

Zboril et al. (1997) analysed properties of the strongly magnetic He-rich star CPD−62 2124 and concluded that its helium abundances might be less reliable due to the presence of strong emission in the Balmer lines. This emission is visible not only in the star but also in the surrounding sky. The authors concluded that the origin of the emission lies in a nebula, but it was not clear whether this nebula surrounds the star or is located in the foreground or in the background. The magnetic field of CPD−62 2124 is dominated by a dipolar component of 21 kG (Hubrig et al. 2017). Such an extraordinarily strong magnetic field has never been detected in any other massive early B-type star.

Hubrig et al. (2020c) reported on the discovery of a longitudinal magnetic field in the WN7 star WR 55 showing hydrogen deficiency and the presence of a corotating interacting region (CIR). The signatures of CIRs are different absorption features of photospheric origin propagating with different accelerations at the same time. They were detected in spectroscopic time series observations for only a few massive stars (e.g., Mullan 1984). It was suggested that CIRs are related to the presence of magnetic bright spots, which are indicators of the presence of a global magnetic field (e.g., Ramiaramanantsoa et al. 2014). Cappa et al. (2009) investigated the distribution of molecular gas related to the ring nebula RCW 78 around WR 55 and concluded that this star is not only responsible for the ionisation of the gas in the nebula, but also for the creation of the interstellar bubble. Their analysis indicated that the star formation in this region is induced by the strong wind of WR 55.

Similar to WR 55, the massive magnetic Of?p star NGC 1624-2 with a dipole strength $\geqslant$ 16 kG (Järvinen et al. 2021) is the main ionising source in the young cluster NGC 1624 associated with the H II region Sh2-212 and showing sequential star formation toward the direction of molecular clumps (Jose et al. 2011).

The results presented above show that surveys of the magnetic fields in massive stars are also of exceptional interest for star formation theories. Indeed, dedicated searches for the presence of magnetic fields in stars located in active sites of star formation led to the detection of rather strong magnetic fields in a few very young massive stars. In the framework of an ESO Large Program, low-resolution spectropolarimetric observations of the three most massive central stars in the Trifid nebula, HD 164492A, HD 164492C, and HD 164492D, were carried out with FORS 2 in 2013 April. These observations revealed the presence of a strong longitudinal magnetic field of about 500–600 G in the poorly studied component HD 164492C (Hubrig et al. 2014a). Follow-up observations showed that HD 164492C is in fact a triple stellar system formed by an eccentric close spectroscopic binary (Ca1-Ca2) with a period of 12.5 days, and a massive early B-type magnetic tertiary Cb (González et al. 2017). A rather strong mean longitudinal magnetic field was also discovered in the early-B star ρ Oph A located in the star-forming cluster LDN 1688 belonging to the ρ Ophiuchus star-forming cloud. The radiation field in LDN 1688 is dominated by two early B-type stars, ρ Oph A (HD 147933) and ρ Oph B (HD 147934). ρ Oph A possesses a magnetic field with a lower limit for the dipole strength of about 2–3 kG (Hubrig et al. 2018; Pillitteri et al. 2018).

The detection of magnetic fields in massive stars located in the youngest star-forming regions implies that these targets may play a pivotal role in understanding the origin of magnetic fields in all massive stars. It is striking that both magnetic massive stars located in young star-forming regions show similar characteristics, such as fast rotation and the presence of X-ray emission. Their stellar surfaces also show helium abundance variations typical for He-rich Bp stars with large-scale organised magnetic fields. Especially intriguing is the presence of an extended blue reflection nebula around the system ρ Oph AB. In parallel with the Of?p star HD 148937, it can be speculated that ρ Oph A can similarly be a merger product and that the surrounding nebula is created by the ejected material. Obviously, it would be important to investigate the chemical composition of the material of the nebula around ρ Oph A to determine its origin. Since star formation in molecular clouds is assumed to be triggered by the dynamical action of winds from massive stars, it is necessary to understand how magnetised winds from magnetic massive stars formed during the first episodes of star formation influence their environments, including nearby sites of star formation and protoplanetary disks surrounding pre-main sequence stars.

3.2 Limitations of the Various Observing and Modelling Techniques

Magnetic field studies relying on individual line profiles require spectropolarimetric observations of very high quality. Such data can currently be obtained only for a small fraction of bright Ap/Bp magnetic stars. Further, magnetic field measurement errors depend not only on the signal-to-noise ratio (S/N), but also on the complexity of the spectrum, i.e., spectral type, $v \sin i$, binarity, etc. For lower quality spectra of the more numerous fainter stars the polarisation signal can be averaged over lines, still revealing clear polarisation signatures in the average Stokes parameter profiles calculated with the least-squares deconvolution method (LSD; Donati et al. 1997, Kochukhov et al. 2010). A more sophisticated method, the SVD method for Stokes profile reconstruction, was introduced by Carroll et al. (2012)—see Section 2.10 for more details on LSD and SVD. The application of the LSD and SVD techniques to high S/N high-resolution polarimetric spectra permits to obtain also spatial and structural information about the field topology. Even if the star is observed close to the magnetic equator, where the longitudinal magnetic field strength is close to 0, it is still possible to notice a typical crossover signature, indicating that this star possesses a magnetic field.

To extract information on stellar surface structures, various Doppler and Zeeman–Doppler imaging (DI and ZDI) techniques are applied using LSD or SVD Stokes profiles. However, as Kochukhov & Wade (2016) demonstrated, a combination of several compounding factors (slow stellar rotation, absence of linear polarisation data, unusually high degree of field complexity) can render a ZDI inversion highly ambiguous. Moreover, according to Alecian & Stift (2017), to achieve analysis results as realistic as possible, it is necessary to carry out 3D mapping, including atomic diffusion that stratifies the abundances of metals in dependence on the local magnetic field strength and the direction with respect to the surface normal.

Usually, the polarised flux of most objects does not exceed 1% of the total flux and the polarisation signal should be measured with an accuracy of at least one-tenth. Obviously, observations with telescopes with apertures between 2 and 3.6 m could not deliver the necessary high *S/N* polarimetric spectra for faint stars and for the majority of massive stars. Especially O-type stars and Wolf–Rayet (WR) stars are rather faint: the Bright Star Catalog contains only about 50 O-type stars and only very few WR stars. Besides, measurements of longitudinal magnetic fields in massive stars are difficult, as only a rather small number of spectral lines compared to the rich spectra of lower-mass Ap and late-Bp stars are available. In addition, spectral lines in massive stars are usually strongly broadened by rapid rotation and macroturbulence, and frequently appear in emission or display P Cygni profiles. In high-resolution spectropolarimetric observations, broad spectral lines frequently extend over adjacent orders, so that it is necessary to adopt order shapes to get the best continuum normalisation. Zeeman signatures in the spectra of magnetic massive stars are generally very noisy and small. Increasing the S/N by increasing the exposure time is frequently limited by the length of the rotation period of the star. In the framework of the MiMeS survey, a median precision of the longitudinal magnetic field measurements, using the LSD technique, of 50 G was achieved for half of the observations, while only 5% of the observations were acquired with $\sigma < 20$ G.

In view of the large line broadening in massive stars, to search for the presence of magnetic fields, low-resolution spectropolarimeters, among them FORS 2 and prior to it FORS 1 on the VLT, appear to be currently the best suitable instruments for this type of research. Importantly, the use of low-resolution spectropolarimeters installed at telescopes with large apertures allows astronomers to get access to surveys of magnetic fields in massive stars even in galaxies in our neighbourhood. Unfortunately, apart from the two FORS 1 and FORS 2 instruments, the low-resolution spectropolarimeter on SALT, and the rarely used spectropolarimetric modes of the Faint Object Camera and Spectrograph at the Subaru Telescope (FOCAS) and of the Gemini Multi-Object Spectrographs (GMOS) at the Gemini Observatory, no further low-resolution spectropolarimetric capabilities are available on any of the 8–10mclass telescopes.

Low-resolution spectropolarimeters are also best suited to search for magnetic fields in very fast rotating B stars with decretion disks (Be stars) with very short rotation periods, massive pulsating β Cephei stars, and slowly pulsating (SPB) stars. High-resolution spectropolarimetric observations of pulsating stars and stars with short rotational periods frequently fail to show credible measurement results if the whole sequence of subexposures at different retarder waveplate angles including overheads has a duration comparable to the timescale of the pulsation or the rotation variability. As an example, even for the bright ($m_V = 4.3$) β Cephei star ξ^1 CMa with a pulsation period of 5 h, a full HARPSpol sequence of subexposures requires about 30 min. In contrast, one FORS 2 observation of the same star lasts less than 10 min, which is relevant to sample the magnetic field over the stellar pulsation period in finer detail. Owing to the strong changes in the line profile positions and the shapes in the spectra of pulsating stars, a method using observations with longer exposure times leads to erroneous wavelength shifts and thus to wrong values for the longitudinal magnetic field (e.g., Järvinen et al. 2017). Similarly, for stars with short rotation periods, high-resolution spectropolarimetric observations lead to phase smearing and a much lower magnetic signal.

Notably, using too low resolution in the spectropolarimetric observations of slowly rotating stars with sharp lines can cause a significant degradation of the measured field strength. A few years ago, using the low-resolution spectropolarimeter FORS 2 attached to the 8 m VLT telescope, Bagnulo et al. (2017) attempted to measure magnetic fields in five Of?p stars in the Magellanic Clouds. Among the polarimetrically studied O-type stars, all five known Galactic stars with Of?p classification possess measurable magnetic fields. These stars exhibit recurrent and apparently periodic spectral variations in Balmer, He I, C III and Si III lines, sharp emission or P Cygni profiles in He I and the Balmer lines, and a strong C III blend in emission around 4650 Å (Walborn 1972). However, no field detection was achieved in any Of?p star identified in the Magellanic Clouds from the obtained polarimetric data sets in the framework of this program. The team used GRISM 1200B, a slit width of 1.0″ and binning of 2 to obtain a spectral resolving power of only ~700. Yet, it is well known that Of?p stars are very slow rotators, implying the presence of rather narrow line profiles in their photospheric spectra. Out of five Of?p stars identified in the Magellanic Clouds, three stars have rotation periods $\gtrsim$400 days (Nazé et al. 2015). However, the use of extremely low spectral resolution unavoidably causes dilution of the magnetic signal preserved in the rather sharp spectral lines. The impact of observations at low resolution on the detectability of magnetic fields in sharp-lined spectra is displayed in Figure 3.3. In this figure Stokes I and Stokes V LSD profiles in HARPSpol spectra ($R = 110,000$) of the strongly magnetic massive early B-type star HD 96446 are overplotted by the same spectra degraded to the resolution of about 10,000. It is clearly visible that line profiles in the spectra observed at lower resolution are much broader and shallower. More importantly, the measured magnetic field strength in the low-resolution spectra becomes weaker by almost a factor of 2.

3.3 Adverse Effects in the Reduction and Magnetic Field Measurement Techniques

In magnetic Ap/Bp stars with large-scale organised, predominantly dipolar magnetic fields, the non-uniformly distributed chemical elements show a certain symmetry with respect to the magnetic field configuration. As an example, rare-earth element spots are frequently identified close to magnetic poles, while iron peak element spots tend to concentrate in the vicinity of the magnetic equator (e.g., Elkin et al. 2010). Obviously, combining spectral lines of various elements may lead to the dilution of the magnetic signal or even to its (partial) cancellation, if enhancements of different elements occur in regions of opposite magnetic polarity. Hence, it is advisable to use in the magnetic field analysis line samples for individual elements separately. To take

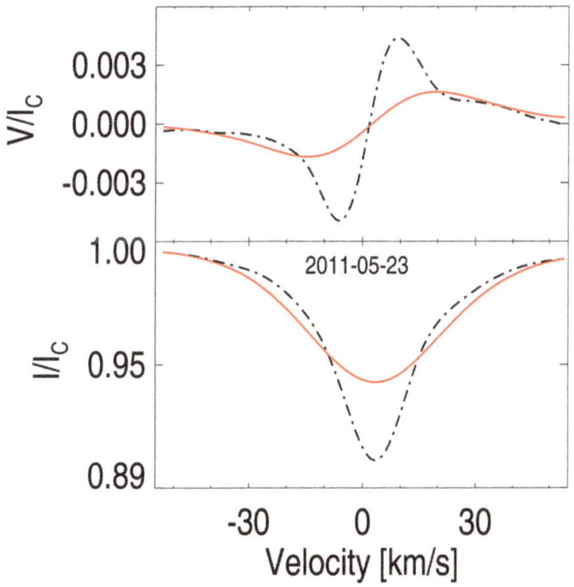

Figure 3.3. Impact of observations at low resolution on the detectability of magnetic fields in sharp-lined spectra: Stokes I (*bottom panel*) and Stokes V (*top panel*) least-square-deconvolution profiles in original HARPSpol spectra ($R = 110,000$) of the strongly magnetic early B-type star HD 96446 (*black dot–dashed line*) are overplotted by the same spectra degraded to a resolution of about 10,000 (*red line*). The line profiles in the spectra obtained at lower resolution are much broader and shallower and the measured magnetic field strength in the low-resolution spectra becomes weaker by almost a factor of 2.

into account that Ap/Bp stars also show a non-uniform vertical distribution of certain elements, it is necessary to determine the magnetic field strengths resulting from measurements of spectral lines belonging to different ionisation stages or lines with different excitation potentials. Such an analysis permits to investigate the relation between the magnetic field strength and orientation on the one hand and vertical element stratification on the other hand (e.g., Järvinen et al. 2020).

In some stars, e.g., in HgMn stars, the presence of tangled magnetic fields was suggested by several authors who detected quadratic magnetic fields using the moment technique (Mathys & Hubrig 1995; Hubrig 1998). Highly structured magnetic field topologies can reduce the Stokes V signatures in spectral line profiles below the detection threshold (e.g., Hubrig et al. 2020a). Furthermore, as was shown by Briquet et al. (2010) and Korhonen et al. (2013), chemical spots on the surface of HgMn stars are not stable and show a temporal evolution of the spot shapes and the element abundances on timescales as short as a few months. Therefore, it is possible that the detectability of magnetic fields strongly depends on the spot position and the element abundance within the spot, especially when observing epochs are separated by long time intervals (Hubrig et al. 2020a).

Special care is also required in the magnetic studies of Herbig Ae/Be stars due to an impact of the circumstellar (CS) matter on the observed line profiles. Järvinen et al. (2015) revealed significant differences between the appearance of Zeeman features in the HARPSpol spectra of the slowly rotating ($P_{\rm rot}$ = 42.076 days; Hubrig et al. 2011a) magnetic Herbig Ae star HD 101412. In the SVD Stokes V profiles obtained at the first three observing epochs, distinct features appear as a second maximum in the red wings of the Zeeman profiles for neutral Fe. Furthermore, these profiles are slightly shifted to the blue by about 4 km s^{-1}. The SVD Stokes I, Stokes V, and diagnostic null N profiles for the Fe I lines are presented in Figure 3.4. For comparison, the Zeeman features observed for the Fe II lines are shown in Figure 3.5. Different appearances of the Zeeman signatures between the Fe I and Fe II line samples are clearly visible. Such a behaviour seems to indicate contamination of the Fe I lines by the surrounding warm CS matter in the form of a wind. This scenario seems to be in agreement with the previous finding of Hubrig et al. (2009b), who reported a strong contamination of UVES spectra of this star. It was later shown that similar distinct features and shifts appear also in the Zeeman profiles

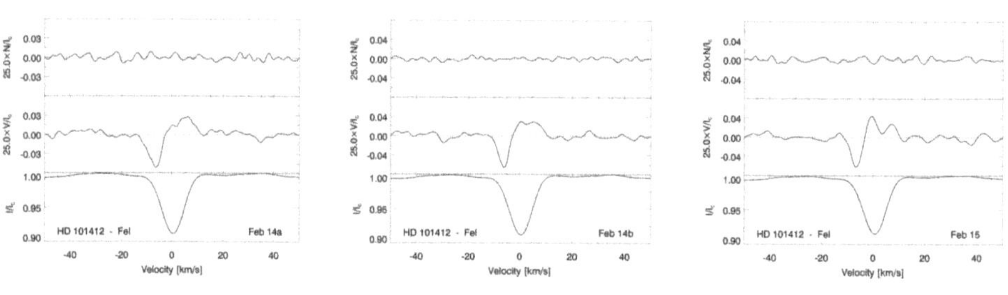

Figure 3.4. The correspondence between SVD Stokes I, Stokes V and diagnostic null N profiles calculated for the Herbig Ae star HD 101412 using a sample of 339 Fe I lines at three different epochs. Credit: Järvinen et al. (2015), reproduced with permission © ESO.

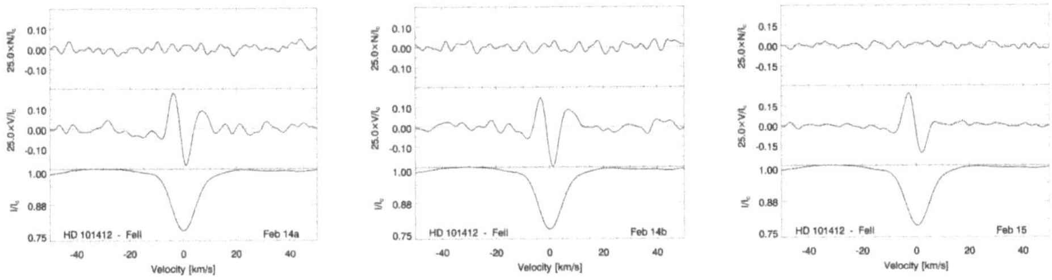

Figure 3.5. The correspondence between SVD Stokes *I*, Stokes *V* and diagnostic null *N* profiles calculated for the Herbig Ae star HD 101412 using a sample of 52 Fe II lines at three different epochs. Credit: Järvinen et al. (2015), reproduced with permission © ESO.

calculated for a sample of Ti II lines (Järvinen et al. 2016). In a circumstellar environment, such as a wind or an accretion disk, lines may form over a relatively large volume and the field topology may be therefore complex not only in latitude and azimuth, but in radius as well. The motion of the CS gas is expected to be governed by the magnetic field, especially in regions of the stellar wind, which is generally assumed to be accelerated by a magnetic centrifuge.

Spectropolarimetric observations of photospheric lines are usually used to determine the geometry of the global magnetic field, while spectropolarimetry of accretion diagnostic lines (e.g., He I λ5876, the Na I doublet, and the Balmer lines) probes the accretion topology in accretion columns, which are perturbed by interactions with the disk. The combination of observations in the visual with those in the near-IR appears most useful, since previous near-IR studies of Herbig Ae/Be stars indicate that the He I λ10830 line is an excellent diagnostic to probe inflow (accretion) and outflow (winds) in the star–disk interaction region (Edwards et al. 2006; Schöller et al. 2016).

It is also of importance to analyse the shapes of the observed Stokes *V* profiles. In some stars, the shape of the Stokes *V* profiles calculated for lines belonging to certain elements displays differences between the areas corresponding to the blue and red parts of these profiles. However, if Stokes *V* profiles are formed in a hydrostatic stellar atmosphere lacking gradients in velocity and magnetic field as a function of optical depth (e.g., Sanchez Almeida et al. 1989; López Ariste 2002), they always have the zero-order moment equal to zero, i.e., the integral of Stokes *V* over the region of the spectral line in Stokes *I* is equal to zero. Differences between the areas corresponding to the blue and red parts of the Stokes *V* profiles were recently detected in several He I emission lines in the spectra of the strongly magnetic Of?p star NGC 1624-2 (Järvinen et al. 2021).

A number of discrepancies in published measurement accuracies has been reported by Bagnulo et al. (2012, 2015), who used the ESO FORS pipeline based on the Reflex toolkit (Izzo et al. 2010) to reduce the full spectropolarimetric stellar content of the FORS 1 archive. As discussed in Section 2.9, for FORS 1/2 observations, the mean longitudinal magnetic field $\langle B_z \rangle$ is defined by the slope of the weighted linear regression through the measured data points, where the weight of each data point is given by the squared *S*/*N* of the Stokes *V* spectrum. The formal 1σ

error of $\langle B_z \rangle$ is obtained from the standard relations for weighted linear regression. This error is inversely proportional to the rms *S*/*N* of Stokes *V*. Bagnulo et al. (2012, 2015) reported the detection of small line shifts in FORS 1 polarimetric spectra and suggested that they are caused by very small instrument flexures, negligible in most of the instrument applications. Further they suggested that the presence of instruments flexures can lead to spurious magnetic field detections and all FORS 1 detections can be considered as reliable only at a significance level greater or equal to 5σ. The assumption that only detections at 5σ significance level are reliable actually means that all published magnetic field detections achieved at a 3σ significance level should be disregarded. Searching for an instrumental or observational systematic origin of the detected line shifts, his team also discussed an impact of seeing conditions and coherence times on the magnetic field measurements in cases when the exposure time is comparable with the atmospheric coherence time (Fossati et al. 2015a). However, in their considerations, it was erroneously assumed that the atmospheric coherence time is in seconds and not in milliseconds. Notably, no other report on the presence of flexures in FORS 1/2 observations was ever published by any instrument user or ESO, so the reason for the line shifts observed by Bagnulo et al. (2012, 2015) was not clear.

Most recently, Hubrig et al. (2020b) discussed this issue concerning line shifts and a related wavelength instability by comparing the FORS 2 spectra extracted using the ESO FORS pipeline and those extracted using a dedicated pipeline written in the MIDAS environment by the previous FORS 1 instrument scientist T. Szeifert. The analysis of underexposed FORS 2 spectra with insufficient S/N showed that the spectral extraction using both pipelines is not working properly and has issues with wavelength stability, causing wavelength shifts of up to about 2 Å (Hubrig et al. 2020b). Surprisingly, the ESO FORS pipeline showed issues with wavelength stability also for spectra obtained at high S/N. The results of a comparison of the extracted spectra using both pipelines, presented in Figure 3.6, indicate that the instabilities in the FORS 1 instrument mentioned by Bagnulo et al. in several papers can be most likely ascribed to an imperfect spectral extraction and wavelength

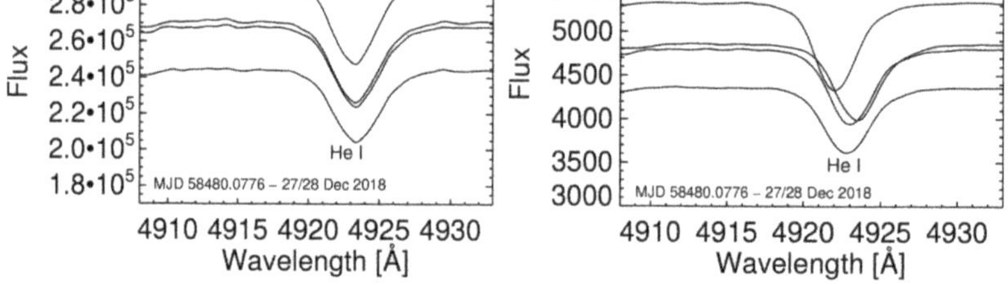

Figure 3.6. Stokes *I* line profiles of He I $\lambda 4922$ over the full sequences of FORS 2 subexposures obtained on a timescale of tens of seconds at MJD 58 487.236 6 at a S/N of 2340 for HD 54879. The spectral extraction was carried out using a pipeline written in the MIDAS environment (*left panel*) and the ESO FORS pipeline (*right panel*). While the MIDAS pipeline shows no wavelength shifts between the different exposures, the ESO FORS pipeline leads to wavelength shifts of up to about 2 Å. The difference in flux between the two pipelines is caused by different ways in handling the absolute flux scale. Credit: Figure 10 from Hubrig et al. (2020b).

instability using the ESO FORS pipeline. Since all FORS 1 measurements presented by Bagnulo et al. (2012, 2015) in the form of catalogues were obtained using the spectra extracted with the ESO FORS pipeline based on the Reflex toolkit, care should be taken when judging the validity of these catalogued magnetic field measurements. Only in the most recent paper related to FORS 2 spectropolarimetric observations of the magnetic O-type star HD 54879, Bagnulo and collaborators decided to measure the position of the telluric O I λ5577 line on each individual 2D wavelength-calibrated FORS 2 frame extracted with the ESO FORS pipeline to roughly correct for the observed line shifts (Wade et al. 2020).

As the measurement accuracies also depend on photon noise, an improper extraction of the spectra, i.e., the use of smaller extraction windows, would explain the larger values for the measurement uncertainties presented in the catalog of Bagnulo et al. (2015). In this work, for the first time, the authors made available the information on the levels of intensity fluxes in frames extracted using the ESO FORS pipeline and reported which spectral regions were used for the magnetic field measurements. The knowledge on the spectral regions selected for the measurements is important as their choice can also have a significant impact on the measurement results (Hubrig et al. 2004). The examination of the levels of intensity fluxes for each subexposure revealed that they are systematically lower, down to 70% in comparison with those obtained in studies of other authors (e.g., Hubrig et al. 2006b; Kurtz et al. 2008; Hubrig et al. 2009a, 2009b), who used for the spectra extraction the pipeline written in the MIDAS environment. Clearly, especially the detection of weak stellar magnetic fields is affected if the extracted fluxes are low. Some examples of flux comparisons measured for a few stars for which magnetic field detections were reported by various authors, but were disregarded in the study of Bagnulo et al. (2015), are presented in Figure 3.7. Further inspection of the S/N values published by Bagnulo et al. (2015) showed that the authors also systematically excluded from the measurements emission lines visible in the spectra of a number of pre-main sequence and massive stars. The reason for this is not clear, as there is no need to differentiate between absorption and emission lines: the used relation between the Stokes V signal and the slope of the spectral line wings holds for both types of lines, so that the signals presented in emission and absorption lines add up rather than cancel. Unfortunately, the fluxes for the left-hand and right-hand polarised spectra

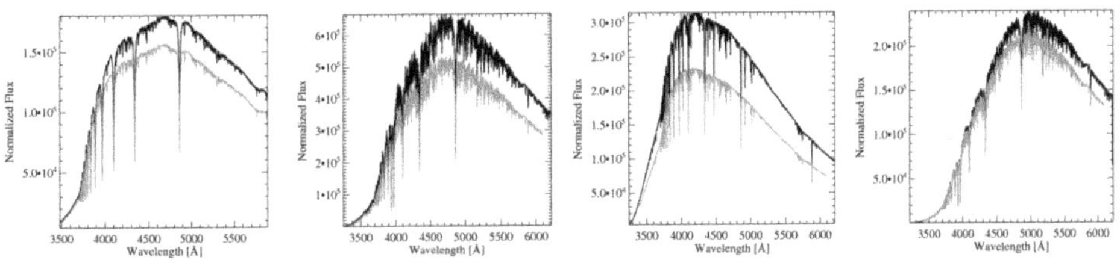

Figure 3.7. Fluxes extracted by Bagnulo et al. (2015; *grey colour*) compared to those using the MIDAS pipeline (*black colour*). The differences in the fluxes are presented (from left to right) for the HgMn star α And (Hubrig et al. 2006b), the δ Scuti star HD 21190 (Kurtz et al. 2008), the nitrogen-rich early B-type star HD 52089 (Hubrig et al. 2009a) and the Herbig Ae star PDS 2 (Hubrig et al. 2015). All these stars were announced in the respective studies as magnetic.

were not made available in the work of Bagnulo et al. (2015). Thus the reproduction of the published measurements is not possible for the readers.

The first detection of the presence of variable slopes in the continua of the *V*/*I* spectra in FORS 1 observations of the X-ray binary Cyg X-1 was reported by Karitskaya et al. (2009, 2010). After rectifying polarimetric FORS 1 spectra, the authors announced the discovery of a weak longitudinal magnetic field at a significance level of $\sim 4\sigma$ in the optical component with spectral type O9.7Iab in this binary system containing the historically first black hole candidate. Variable continuum slopes of *V*/*I* spectra are frequently detected in measurements of different types of stars: in Herbig Ae/Be stars with a complex circumstellar environment, in WR stars with strong dense winds, in O-type stars surrounded by circumstellar nebulae, in A-type supergiants, in central stars of planetary nebulae, etc. Usually, before the calculation of linear regressions, a rectification of the *V*/*I* spectra is performed by fitting and subtracting a line to the data. As an example, the rectified and non-rectified *V*/*I* spectra of the A-type supergiant HD 92207 observed with FORS 2 in 2011 May are presented in Figure 3.8. The impact of rectification is usually not very strong, but important in the surveys of weak magnetic fields: for the A-type supergiant HD 92207 the difference between the mean longitudinal magnetic field obtained in rectified spectra and that obtained in non-rectified spectra is only

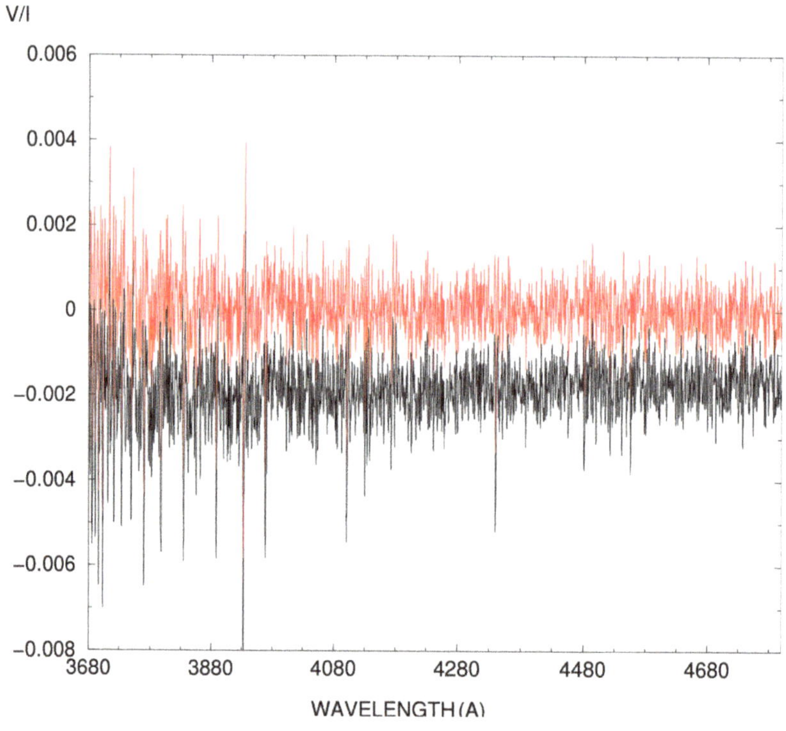

Figure 3.8. Rectified (*red*) and non-rectified (*black*) Stokes *V*/*I* spectra of the A-type supergiant HD 92207 observed in 2011 May. Generally, the Stokes *V* spectrum should not be offset from zero. Credit: Figure 1 from Hubrig et al. (2014b).

about 15 G. Even for the highly linearly polarised object Cyg X–1, Karitskaya et al. (2009) report only a difference between the rectified and the non-rectified FORS 1 Stokes *V* spectra of less than 20 G.

Starting in 2010, Monte-Carlo bootstrapping tests were implemented in the analysis of FORS 2 measurements (Rivinius et al. 2010; Steffen et al. 2014; Hubrig et al. 2014b). They are most often applied with the purpose of deriving robust estimates of standard errors. In these tests, typically 250,000 samples are generated that have the same size as the original data set and the resulting distribution of the $P(\langle B_z \rangle)$ determined from all these newly generated data sets is analysed. Examples of the resulting distributions, all showing smooth Gaussian profiles, are presented in Figure 3.9. Mean and standard deviation of this distribution are identified with the most likely mean longitudinal magnetic field and its 1σ error, respectively. The main advantage of this method is that it provides an independent error estimate.

Obviously, to increase the reliability of magnetic field detections, but also to carry out a quantitative atmospheric analysis and to probe spectral variability, it is advisable to follow-up low spectral resolution detections with high-resolution observations. Subsequent high-resolution spectropolarimetric observations permit in addition to carry out a quantitative atmospheric analysis and to probe spectral variability of stars with detected magnetic fields. The only collaboration that used FORS 2 and HARPS to monitor magnetic fields was the BOB (B-fields in OB stars) collaboration (Morel et al. 2014), which focused on the search of magnetic fields in early B- and O-type massive stars. Combining observations with different instruments allowed the BOB collaboration to report during the last couple of years the presence of magnetic fields in a number of massive stars. As an example, the first detection of a magnetic field in the single slowly rotating O9.7V star HD 54879 was achieved with FORS 2. The follow-up HARPS observations confirmed the presence of a rather strong magnetic field and showed that HD 54879 has a stable and normal optical spectrum (Castro et al. 2015). Monitoring of the spectral and magnetic field variability of this star using FORS 2 and ESPaDOnS spectropolarimetric observations over the last years unveiled distinct changes of the magnetic field strength and of the intensity of the Hα line indicating that its rotation period is extremely long, at least several years (e.g., Hubrig et al. 2020b). Within the BOB collaboration, for the first time, extremely weak (<30 G in absolute value) longitudinal magnetic fields were detected in the massive early B-type stars β CMa and ϵ CMa (Fossati et al. 2015b). Both stars were previously detected as magnetic by Hubrig et al. (2006a, 2009a) using FORS 1 observations.

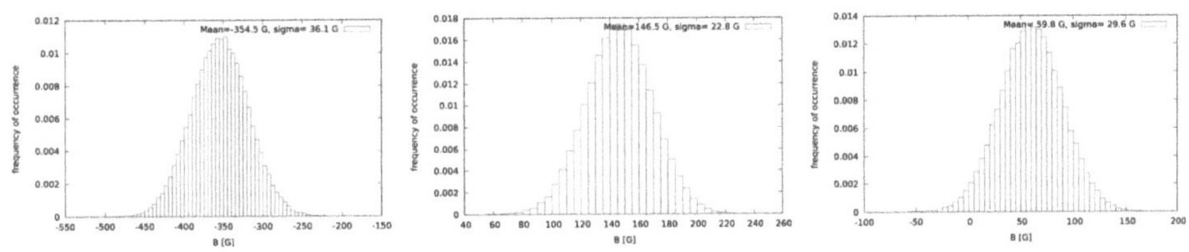

Figure 3.9. Distributions of Monte-Carlo bootstrapping tests for the data sets of the A-type supergiant HD 92207 from three different epochs. The widths of the presented profiles are used to determine the uncertainties of the FORS 2 $\langle B_Z \rangle$ measurements. Credit: Figure 10 from Hubrig et al. (2014b).

References

Alecian, G., & Stift, M. J. 2017, MNRAS, 468, 1023
Bagnulo, S., Landstreet, J. D., Fossati, L., & Kochukhov, O. 2012, A&A, 538, A129
Bagnulo, S., Fossati, L., Landstreet, J. D., & Izzo, C. 2015, A&A, 583, A115
Bagnulo, S., Nazé, Y., Howarth, I. D., et al. 2017, A&A, 601, A136
Bieging, J. H., Abbott, D. C., & Churchwell, E. B. 1989, ApJ, 340, 518
Bohlender, D. 2005, in EAS Publications Ser. 17, ed. G. Alecian, O. Richard, & S. Vauclair, 83
Briquet, M., Korhonen, H., González, J. F., Hubrig, S., & Hackman, T. 2010, A&A, 511, A71
Cappa, C. E., Rubio, M., Martín, M. C., & Romero, G. A. 2009, A&A, 508, 759
Carroll, T. A., Strassmeier, K. G., Rice, J. B., & Künstler, A. 2012, A&A, 548, A95
Castelli, F., & Hubrig, S. 2007, A&A, 475, 1041
Castro, N., Fossati, L., Hubrig, S., et al. 2015, A&A, 581, A81
Catanzaro, G., Giarrusso, M., Leone, F., et al. 2016, MNRAS, 460, 1999
Chlebowski, T. 1989, ApJ, 342, 1091
Cohen, D. H., de Messières, G. E., MacFarlane, J. J., et al. 2003, ApJ, 586, 495
Cowley, C. R., Hubrig, S., Ryabchikova, T. A., et al. 2001, A&A, 367, 939
Donati, J. F., Semel, M., Carter, B. D., Rees, D. E., & Collier Cameron, A. 1997, MNRAS, 291, 658
Donati, J. F., Babel, J., Harries, T. J., et al. 2002, MNRAS, 333, 55
Donati, J. F., Howarth, I. D., Bouret, J. C., et al. 2006, MNRAS, 365, L6
Edwards, S., Fischer, W., Hillenbrand, L., & Kwan, J. 2006, ApJ, 646, 319
Elkin, V. G., Mathys, G., Kurtz, D. W., Hubrig, S., & Freyhammer, L. M. 2010, MNRAS, 402, 1883
Ferrario, L., Pringle, J. E., Tout, C. A., & Wickramasinghe, D. T. 2009, MNRAS, 400, L71
Fossati, L., Bagnulo, S., Landstreet, J. D., & Kochukhov, O. 2015a, in ASP Conf. Ser. 494 Physics and Evolution of Magnetic and Related Stars, ed. Y. Y. Fossati, I. I. Romanyuk, & D. O. Kudryavtsev (San Francisco, CA: ASP), 63
Fossati, L., Castro, N., Morel, T., et al. 2015b, A&A, 574, A20
González, J. F., Hubrig, S., Przybilla, N., et al. 2017, MNRAS, 467, 437
Henrichs, H. F., de Jong, J. A., Donati, J. F., Smith, M. A., Henrichs, H. F., Fabregat, J., et al. 2000, in ASP Conf. Ser. 214, IAU Colloq. 175: The Be Phenomenon in Early-Type Stars (San Francisco, CA: ASP), 324
Hubrig, S., Kurtz, D. W., Bagnulo, S., et al. 2004, A&A, 415, 661
Hubrig, S., Briquet, M., Schöller, M., et al. 2006a, MNRAS, 369, L61
Hubrig, S., North, P., Schöller, M., & Mathys, G. 2006b, AN, 327, 289
Hubrig, S., Briquet, M., Morel, T., et al. 2008, A&A, 488, 287
Hubrig, S., Briquet, M., De Cat, P., et al. 2009a, AN, 330, 317
Hubrig, S., Stelzer, B., Schöller, M., et al. 2009b, A&A, 502, 283
Hubrig, S., Mikulášek, Z., González, J. F., et al. 2011a, A&A, 525, L4
Hubrig, S., Schöller, M., Kharchenko, N. V., et al. 2011b, A&A, 528, A151
Hubrig, S., Schöller, M., Ilyin, I., et al. 2013, A&A, 551, A33
Hubrig, S., Fossati, L., Carroll, T. A., et al. 2014a, A&A, 564, L10
Hubrig, S., Schöller, M., & Kholtygin, A. F. 2014b, MNRAS, 440, 1779
Hubrig, S., Carroll, T. A., Scholler, M., & Ilyin, I. 2015, MNRAS, 449, L118
Hubrig, S., Mikulášek, Z., Kholtygin, A. F., et al. 2017, MNRAS, 472, 400
Hubrig, S., Schöller, M., Järvinen, S. P., et al. 2018, AN, 339, 72

Hubrig, S., Järvinen, S. P., Korhonen, H., et al. 2020a, MNRAS, 495, L97
Hubrig, S., Järvinen, S. P., Schöller, M., & Hummel, C. A. 2020b, MNRAS, 491, 281
Hubrig, S., Schöller, M., Cikota, A., & Järvinen, S. P. 2020c, MNRAS, 499, L116
Hubrig, S. 1998, CoSka, 27, 296
Hubrig, S. 2008, RMxAC, 33, 26
Izzo, C., de Bilbao, L., Larsen, J., et al. 2010, Proc. SPIE, 7737, 773729
Järvinen, S. P., Carroll, T. A., Hubrig, S., et al. 2015, A&A, 584, A15
Järvinen, S. P., Hubrig, S., Schöller, M., et al. 2016, AN, 337, 329
Järvinen, S. P., Hubrig, S., Ilyin, I., Schöller, M., & Briquet, M. 2017, MNRAS, 464, L85
Järvinen, S. P., Hubrig, S., Mathys, G., et al. 2020, MNRAS, 499, 2734
Järvinen, S. P., Hubrig, S., Schöller, M., et al. 2021, MNRAS, 501, 4534
Jose, J., Pandey, A. K., Ogura, K., et al. 2011, MNRAS, 411, 2530
Karitskaya, E. A., Bochkarev, N. G., Hubrig, S., et al. 2009, arXiv: 0908.2719
Karitskaya, E. A., Bochkarev, N. G., Hubrig, S., et al. 2010, IBVS, 5950, 1
Kochukhov, O., & Wade, G. A. 2016, A&A, 586, A30
Kochukhov, O., Makaganiuk, V., & Piskunov, N. 2010, A&A, 524, A5
Korhonen, H., González, J. F., Briquet, M., et al. 2013, A&A, 553, A27
Kurtz, D. W., Hubrig, S., González, J. F., van Wyk, F., & Martinez, P. 2008, MNRAS, 386, 1750
Kurtz, D. W. 1982, MNRAS, 200, 807
López Ariste, A. 2002, ApJ, 564, 379
Leitherer, C., & Chavarria-K, C. 1987, A&A, 175, 208
Mahy, L., Hutsemékers, D., Nazé, Y., et al. 2017, A&A, 599, A61
Mathys, G., & Hubrig, S. 1995, A&A, 293, 810
Mathys, G. 2004, in IAU Symp. 215, Stellar Rotation ed. A. Maeder, & P. Eenens (San Francisco, CA: ASP), 270
Mathys, G. 2017, A&A, 601, A14
Morel, T., Butler, K., Aerts, C., Neiner, C., & Briquet, M. 2006, A&A, 457, 651
Morel, T., Hubrig, S., & Briquet, M. 2008, A&A, 481, 453
Morel, T., Castro, N., Fossati, L., et al. 2014, Msngr, 157, 27
Mullan, D. J. 1984, ApJ, 283, 303
Nazé, Y., Rauw, G., Pollock, A. M. T., Walborn, N. R., & Howarth, I. D. 2007, MNRAS, 375, 145
Nazé, Y., Walborn, N. R., Rauw, G., et al. 2008, AJ, 135, 1946
Nazé, Y., ud-Doula, A., Spano, M., et al. 2010, A&A, 520, A59
Nazé, Y., Bagnulo, S., Petit, V., et al. 2012, MNRAS, 423, 3413
Nazé, Y., Walborn, N. R., Morrell, N., Wade, G. A., & Szymański, M. K. 2015, A&A, 577, A107
Nesvacil, N., Lüftinger, T., Shulyak, D., et al. 2012, A&A, 537, A151
Netopil, M., Paunzen, E., Hümmerich, S., & Bernhard, K. 2017, MNRAS, 468, 2745
Pillitteri, I., Fossati, L., Castro Rodriguez, N., Oskinova, L., & Wolk, S. J. 2018, A&A, 610, L3
Preston, G. W. 1974, ARA&A, 12, 257
Przybylski, A. 1961, Natur, 189, 739
Ramiaramanantsoa, T., Moffat, A. F. J., Chené, A.-N., et al. 2014, MNRAS, 441, 910
Reiners, A., Stahl, O., Wolf, B., Kaufer, A., & Rivinius, T. 2000, A&A, 363, 585
Rivinius, T., Szeifert, T., Barrera, L., et al. 2010, MNRAS, 405, L46
Ryabchikova, T., Piskunov, N., Kochukhov, O., et al. 2002, A&A, 384, 545
Sadakane, K., & Nishimura, M. 2018, PASJ, 70, 40

Sanchez Almeida, J., Collados, M., & del Toro Iniesta, J. C. 1989, A&A, 222, 311
Schöller, M., Pogodin, M. A., Cahuasquí, J. A., et al. 2016, A&A, 592, A50
Schulz, N. S., Canizares, C. R., Huenemoerder, D., & Lee, J. C. 2000, ApJL, 545, L135
Steffen, M., Hubrig, S., Todt, H., et al. 2014, A&A, 570, A88
Tout, C. A., Wickramasinghe, D. T., Liebert, J., Ferrario, L., & Pringle, J. E. 2008, MNRAS, 387, 897
Tutukov, A. V., & Fedorova, A. V. 2010, ARep, 54, 156
Wade, G. A., Bagnulo, S., Keszthelyi, Z., et al. 2020, MNRAS, 492, L1
Walborn, N. R. 1972, ApJL, 176, L119
Zboril, M., North, P., Glagolevskij, Y. V., & Betrix, F. 1997, A&A, 324, 949

Swetlana Hubrig and Markus Schöller

Chapter 4

Ap/Bp Stars

The chemically peculiar Ap and late-type Bp stars, which will in the following be called the Ap/Bp stars, play a key role in the efforts to understand relevant stellar physics, since it is in these stars that the effects of the various processes acting below, in and above the stellar atmospheres are most clearly visible. Ap/Bp stars generally possess large-scale organised magnetic fields that can be diagnosed through observations of circular polarisation in spectral lines. However, in spite of the overall progress that has been made in the studies of Ap/Bp stars over the last decades, the fundamental questions, how their magnetic fields are generated and why they constitute only a fraction of the intermediate-mass main-sequence stars remain unanswered.

4.1 General Background

A large variety of the physical processes that occur in the atmospheres of upper main sequence stars have not yet been fully incorporated into stellar models, or even securely identified and understood. These processes include convection, turbulence, meridional circulation currents, diffusion of trace elements within the dominant hydrogen plasma and mass loss through stellar winds. The magnetic Ap/Bp stars constitute the best studied group of early-type magnetic stars and represent a unique laboratory to study various physical effects of magnetic fields in stellar atmospheres. In the spectra of these stars, lines of some elements are abnormally strong (e.g., Si, Sr, rare earths) or weak (in particular, He). They undergo periodic variations of magnitude (in various photometric bands) and spectral line intensities and the known periods of variability range from half a day to several decades. It is generally assumed that the spectral variability of this type of stars is intimately related to the presence of large-scale organised magnetic fields in their atmospheres. Surface properties of Ap/Bp stars are non-uniform and non-symmetric with respect to their rotation axes and are, by and large, determined by the magnetic field. Magnetic Ap/Bp stars constitute roughly 10%–15% of the intermediate-mass and massive

doi:10.1088/2514-3433/abefccch4

main-sequence stars with spectral types between approximately B8 and F0. For a long time, Ap/Bp stars were the only non-degenerate stars besides the Sun in which magnetic fields had been definitely detected and measured.

3652 Ap/Bp stars, including massive early-type B He-rich stars, are mentioned in the catalogue of Ap, HgMn, and Am stars by Renson & Manfroid (2009). The number of Ap/Bp stars has gradually increased over recent years through the exploitation of results from ground based spectroscopic surveys, such as the Apache Point Observatory Galaxy Evolution Experiment (APOGEE; Majewski et al. 2017) and the Large Sky Area Multi-Object Fibre Spectroscopic Telescope (LAMOST; Luo et al. 2012), but also through various photometric surveys, such as ASAS-3 (Bernhard et al. 2015a; Hümmerich et al. 2016), SuperWASP (Bernhard et al. 2015b) and space missions such as STEREO (Wraight et al. 2012), Kepler (Hümmerich et al. 2018), and TESS (Sikora et al. 2019a), for which data acquisition and exploitation of the results is still on-going. Most recently, Hümmerich et al. (2020) identified about one thousand peculiar A and B stars using spectra collected by LAMOST.

Variable magnetic fields are most frequently diagnosed through the mean longitudinal magnetic field and the mean magnetic field modulus. The mean longitudinal magnetic field $\langle B_z \rangle$ is the line-intensity weighted average over the visible stellar hemisphere of the component of the magnetic field along the line of sight. $\langle B_z \rangle$ is strongly dependent on the line of sight and can be obtained for Ap/Bp stars regardless of their rotation rates. The observed strengths of detected longitudinal fields range from a few tens to tens of thousands of Gauss. The star with the strongest magnetic field currently known, Babcock's star (HD 215441), is a B8V star with a surface dipole field strength of 34 kG (Babcock 1960; Preston 1969). To first order, the magnetic fields of the vast majority of the Ap/Bp stars are dipole-like, with the magnetic axis inclined to the rotation axis.

In stars that have a sufficiently strong magnetic field and that rotate slowly, it is possible to measure the mean magnetic field modulus $\langle B \rangle$, which is the line-intensity weighted average over the visible stellar hemisphere of the modulus of the magnetic field. It is determined from measurement of the separation of the π and σ, or σ^+ and σ^- components of lines that show resolved magnetic splitting. Although to some degree dependent on the geometry of the observation, $\langle B \rangle$ has the advantage of being primarily determined by the intrinsic stellar magnetic field strength. As of today, 84 Ap stars with magnetically split lines have been identified in optical data (Mathys 2017). Very recently, resolved magnetically split lines were detected in 157 Ap/Bp stars using near-infrared H-band spectra provided by the APOGEE survey (Chojnowski et al. 2019). This study clearly showed the advantage of using longer wavelengths to search for stars with resolved magnetically split lines by increasing the number of Ap/Bp stars with measured $\langle B \rangle$ by 183% and presented an extension of high-resolution Ap/Bp star observations to fainter magnitudes. Spectroscopy at longer wavelengths provides several advantages in terms of resolving magnetically split lines. The wavelength separation of magnetically split components is proportional to the square of the wavelength, while the Doppler effect has a linear dependence on wavelength. Hence, for a fixed effective Landé factor, the spectral resolution required to resolve the split components of a line can be significantly lower in the H-band than in the optical. The lines most frequently resolved into their split components for these stars in the H-band primarily pertained to Ce III, Cr II, Fe I, Mn II, Si I, and Ca II, in addition to one or several unidentified ions. Full APOGEE spectra of one narrow-lined non-magnetic star (2MASSJ 01023033+7438487) along with a strongly magnetic Ap star for which the unknown lines are magnetically split (2MASSJ 06010117+3214538) are displayed in Figure 4.1 (Chojnowski et al. 2019).

Studies of stars with resolved magnetically split lines indicate that magnetic fields of Ap/Bp stars cover the whole stellar surface (e.g., Mathys et al. 1997). They have a significant degree of large-scale organisation, such that the range of field strengths over the stellar surface is relatively narrow. The existence of a cut-off at the low end of the distribution of the phase-averaged mean magnetic field moduli of Ap/Bp stars with resolved magnetically split lines, at about 2.8 kG, was discussed by Mathys et al. (1997) and later confirmed by Mathys (2017). Mathys (2017) suggested that this cut-off at the low end of the distribution reflects the probable existence of a gap in the distribution of the magnetic field strengths in slowly rotating Ap stars, below which there is a separate population of stars with fields weaker than $\sim$2 kG.

Based on longitudinal field measurements of a small sample of 28 magnetic Ap stars with poorly constrained magnetic field strengths, Aurière et al. (2007) concluded that there exists a critical dipole field strength, $B_{\rm d} \approx 300$ G, which corresponds to the minimum field strength for a star to maintain the stability of its magnetic field. Only two stars in the sample of 28 Ap stars showed a dipole strength below 300 G. On the other hand, spectropolarimetric observations presented by other authors indicate that a number of stars possess very weak longitudinal magnetic fields (e.g., Ryabchikova et al. 2004; Alecian et al. 2016). It was also shown that although some stars have very weak or even undetectable longitudinal magnetic fields, they can possess nonzero mean quadratic magnetic fields (e.g., Mathys & Hubrig 2006), which are determined from the study of the second-order moment of the line profiles recorded in unpolarised light. A number of studies are sensitive to weak longitudinal magnetic fields, down to a few tens of Gauss (e.g., Shorlin et al. 2002; Hubrig et al. 2006). However such weak fields are detectable only provided that they have enough large-scale organisation so that contributions of various parts of the stellar surface do not cancel out in spectropolarimetric observations, and that the geometry of the observation (i.e., the relative orientations of the magnetic field and of the line of sight) is favourable. Notably, measurements of the magnetic quadratic field are not affected by such restrictions. Obviously, a more representative study based on a larger sample of Ap/Bp stars is needed to prove whether a critical value for the stability of a large-scale magnetic field exists. Such information is especially important for the theoretical understanding of the origin of magnetic Ap/Bp stars.

Studying merely the variation phase curve of the mean longitudinal magnetic field over the rotation period, which usually does not significantly depart from a sinusoid, no model more complex than a single dipole, with its axis inclined to the stellar rotation axis can be constrained. To fully describe the magnetic field geometry, all four Stokes parameters have to be observed. As linear polarisation is a second-order effect in the field strength (e.g., Mathys 2002), considerably higher signal-to-noise is

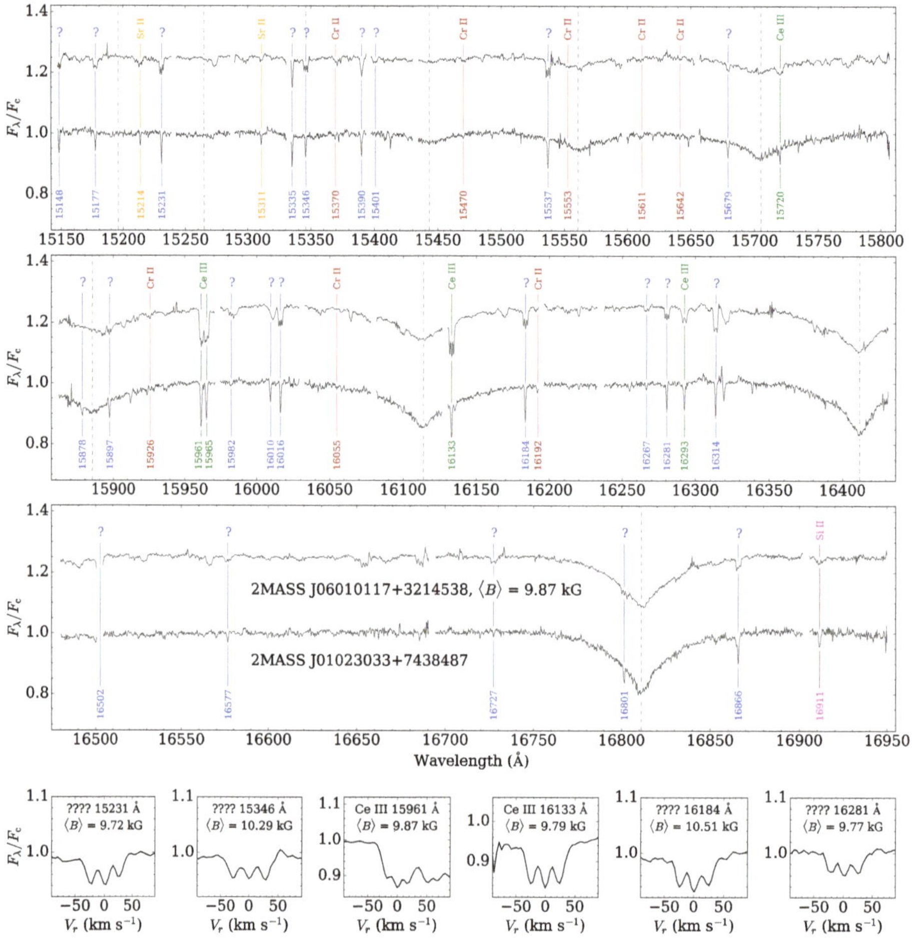

Figure 4.1. *Top three panels:* APOGEE spectra of 2MASS J06010117+3214538 (*top*) and 2MASS J01023033 +7438487 (*bottom*), two stars exhibiting unidentified lines, which are magnetically split in the first star. Blue vertical lines indicate the unknown lines (for which the labels are the vacuum rest wavelengths in Å) and green lines indicate Ce III lines. A few other species in common between the two stars are labelled. Strong airglow lines have been masked from the spectra, including one that typically distorts the profile of the unknown line at 16502 Å. *Bottom panel:* a closer view of magnetically split lines in the spectrum of 2MASS J06010117 +3214538. The feature redwards of Ce III λ15961 is a blend of Si I λ15964 and Ce III λ15965. The mean magnetic field modulus obtained from the measured splitting of each line is given. Credit: Figure 1 from Chojnowski et al. (2019), © 2019. The American Astronomical Society. All rights reserved.

required for the detection and characterisation of line profiles in linearly polarised light than in circularly polarised light. Using observations of the magnetic Ap star 53 Cam in all four Stokes parameters and the Magnetic Doppler Imaging technique (MDI), Kochukhov et al. (2004) reported that the structure of the magnetic field of 53 Cam is much more complex than any low-order multipole expansion. Noteworthy, most of the previously reported studies employing all four Stokes parameters were restricted to mapping the horizontal element abundance and the strength and orientation of the magnetic vector distribution. More recent studies however showed that self-consistent modelling has to be three-dimensional and take into account vertical abundance stratifications arising from atomic diffusion and the magnetic field gradient. Already twenty years ago, Cowley et al. (2001) reported on the existence of a core-wing anomaly in hydrogen Balmer lines of a number of magnetic Ap/Bp stars, indicating that the structure of the atmosphere in magnetic stars cannot be represented by standard model atmospheres. A number of stars also have field geometries that are definitely non-axisymmetric and are characterised by dipoles that are offset from the centre of the star both in the direction of the magnetic axis and perpendicular to it. A few years ago, Alecian & Stift (2017) presented for the first time a three-dimensional model with abundance stratifications arising from atomic diffusion of 16 metals, adopting a non-axisymmetric magnetic field geometry inspired by a configuration derived for the strongly magnetic Ap star HD 154708. They found that the chemical elements are distributed in complex patterns in all three dimensions.

4.2 Abundances

Ap/Bp stars show abnormal line strengths that correspond to non-solar element abundances. In some stars, abundance anomalies (with respect to the cosmic abundance pattern) of certain elements may reach up to 5 or 6 orders of magnitude. The observed abundance anomalies are generally explained by the interplay of selective processes (radiative levitation, gravitational settling) operating in the calm radiative atmospheres of these stars (e.g., Richer et al. 2000). Besides the existence of horizontal abundance inhomogeneities over the stellar surface, various observational results point to the existence of vertical abundance gradients. For several Ap stars with strong magnetic fields, a study of Stokes I and V profiles of individual lines belonging to different elements permits to retrieve information not only on the magnetic field geometry, but also on the vertical and horizontal distribution of various chemical elements as a function of the magnetic field structure. As a pioneering work, Hubrig et al. (2018) analysed high-quality polarimetric spectra obtained with HARPSpol for Przybylski's star (HD 101065), which is characterised by very slowly rotation with a probable period of 188 yr and a $\langle B_z \rangle$ of several hundred Gauss. The authors used longitudinal magnetic field measurements carried out using samples of lines belonging to different elements to discuss the anomalous structure of the atmosphere, in particular the inhomogeneous vertical distribution of several chemical elements. Anomalies related to the vertical abundance stratification of the elements Pr and Nd were also for the first time used to establish the presence of a radial magnetic field gradient. However, because Przybylski's star rotates extremely slow, the structure of its magnetic field in the study of Hubrig et al. (2018) remained undefined.

Very recently, Mathys et al. (2020) presented the first accurate determination of the rotation period of the superslowly rotating rapidly oscillating Ap star HD 166473 with $P_{\rm rot} = 3836 \pm 30$ days from the analysis of 56 measurements of the mean magnetic field modulus $\langle B \rangle$ and 21 determinations of the mean longitudinal magnetic field $\langle B_z \rangle$. The latter were based on high- and medium-resolution spectra acquired between 1992 and 2019 at various observatories and with various instrumental configurations, among them high-resolution ESPaDOnS and HARPSpol polarimetric spectra. Due to the very slow rotation and the presence of a strong magnetic field, numerous lines in the spectra of HD 166473 show Zeeman splitting corresponding to a mean magnetic field modulus in the range from 6.4 to 8.6 kG. As already mentioned in Chapter 3, magnetic Ap stars are known to frequently display horizontal abundance inhomogeneities across their surfaces and the spectral lines of elements that have different distributions across the stellar surface thus sample the magnetic field topology in different manners. As an example, Stokes *I* profiles of individual lines belonging to different elements and the corresponding Stokes *V* profiles observed in the spectra of HD 166473 are presented in Figure 4.2. The observed complex structure of the Stokes *V* profiles of various ions can be attributed to the presence of magnetic fields of mixed polarities over the hemisphere of this star that is facing the observer around the phase of the negative extremum of $\langle B_z \rangle$ (Järvinen et al. 2020). By contrast, close to phase 0.6, the Stokes *V* line profiles are mostly S-shaped. Hence, the polarity of the magnetic field is expected to

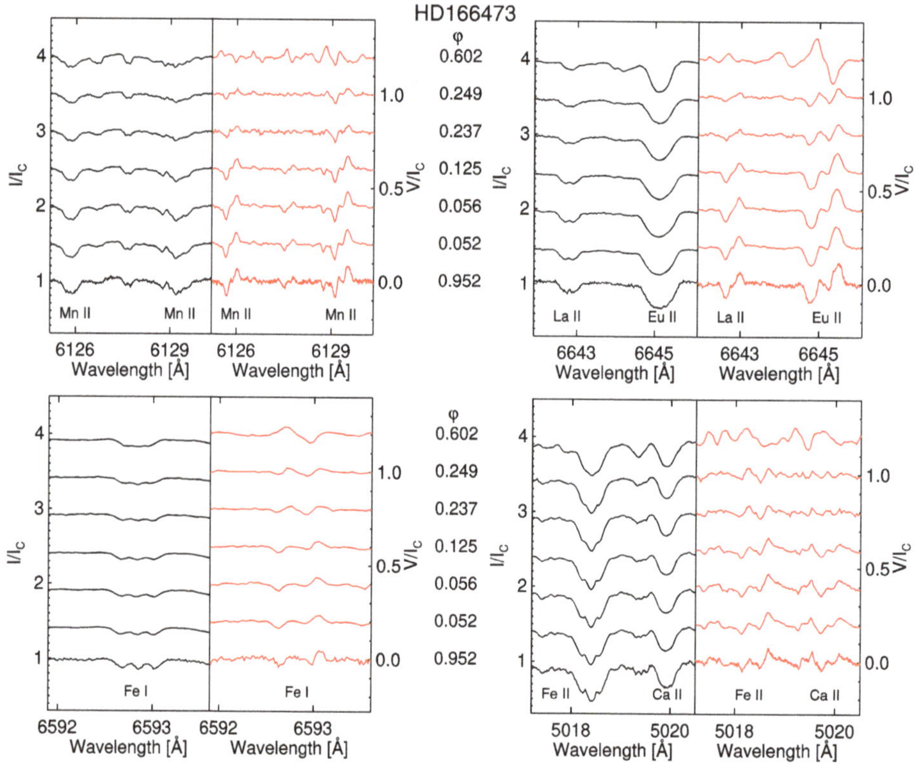

Figure 4.2. Each panel shows the variability of Stokes *I* (left) and Stokes *V* (right) line profiles belonging to a few elements in different ionisation stages over the rotation cycle of HD 166473. *Top left panel:* two Mn II lines; *top right panel:* La II and Eu II lines; *bottom left panel:* Fe I line; *bottom right panel:* Fe II and Ca II lines. Credit: Figure 7 from Järvinen et al. (2020).

be predominantly positive over most of the half of the stellar surface that is visible around the phase of the positive extremum of $\langle B_z \rangle$. To test the vertical abundance stratification of iron peak and rare earth elements, Järvinen et al. (2020) analysed their calculated least-squares deconvolution (LSD) Stokes *I* and *V* profiles. Magnetic field measurements were carried out using spectral lines of elements belonging to neutral and ionised stages. The authors found a clear indication of the existence of a correspondence between the magnetic field strength, its orientation and vertical element stratification: magnetic field values obtained for elements in different ionisation stages close to the magnetic equator were rather similar, whereas the dispersion in field strengths was remarkably large in regions close to the magnetic field poles.

The mean longitudinal magnetic field values for different elements phased with the rotation period of 3836 days are presented in Figure 4.3. In general, the absolute values of the longitudinal field that were determined using all rare earth element line lists were significantly larger close to the phases of the two extrema than the absolute values that were obtained using lines belonging to the Fe and Cr line lists. The largest absolute values of the mean longitudinal magnetic field at the rotation phases corresponding to the negative extremum, up to −4160 ± 226 G, were derived using the La II line list, followed by the measurements using the Pr III line list with $\langle B_z \rangle = -3740 \pm 343$ G and the Ce II line list with $\langle B_z \rangle = -3372 \pm 247$ G. The largest

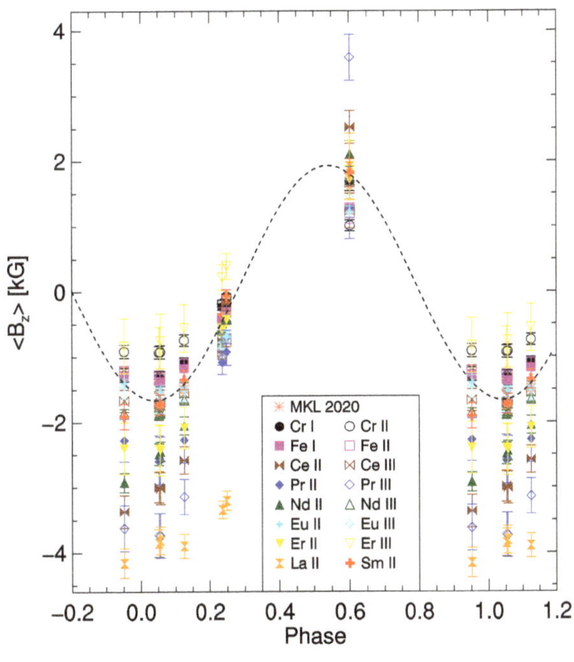

Figure 4.3. Mean longitudinal magnetic field $\langle B_z \rangle$ values of HD 166473 calculated for various line samples. Additionally, the $\langle B_z \rangle$ values from Mathys et al. (2020) measured using exclusively Fe I lines (identified as MKL 2020) are presented by asterisks and their best-fit solution for the magnetic phase curve by the dashed line. Credit: Figure 1 from Järvinen et al. (2020).

positive value of the mean longitudinal magnetic field, up to 3584 ± 354 G, was derived using the Pr III line list, followed by the measurement using the Ce II line list, with $\langle B_z \rangle = 2517 \pm 249$ G.

Alecian & Stift (2017) demonstrated that in order to properly approximate the observational profile of the vertical stratification of the element abundances in Ap/Bp stars using theoretical models of time-dependent atomic diffusion, one also needs to introduce a certain mass loss rate. Clearly, future detailed analyses of the vertical stratification of element abundances in the magnetically controlled atmospheres of a representative sample of magnetic Ap/Bp stars will contribute to verifying observationally the impact of stellar rotation and magnetism on the time-dependent atomic diffusion and improve the current understanding of this process. Also the observational profiles of vertical abundance stratification derived for different chemical elements in the atmospheres of Ap/Bp stars compiled in the framework of project VeSElkA (Vertical Stratification of Element Abundances; Khalack & LeBlanc 2015a, 2015b) will help to constrain the theoretical models of the diffusion process.

Not so much is known about the evolutionary changes of chemical abundances of magnetic Ap/Bp stars across the main sequence of the H-R diagram. Bailey et al. (2014) reported that in magnetic late-type Bp stars certain elements, such as the iron-peak elements Cr, Fe, and Ti, decrease in abundance with age, approaching near-solar values close to the terminal-age main sequence, while the abundance of He increases. Unfortunately, for all 15 stars studied, the authors assumed a uniform horizontal and vertical atmospheric abundance distribution. However, as the spectral line profiles in magnetic stars are affected by inhomogeneous distributions of various chemical elements, the corresponding chemical structures in horizontal and vertical direction modulate the observed line profiles. The Zeeman effect also produces a non-negligible impact on the appearance of the intensity profiles of spectral lines, comparable to the variations caused by chemical spots. Therefore, to carefully investigate the trend of the element abundances with evolutionary age, a 3D magnetic Doppler imaging analysis involving simultaneous mapping of the magnetic field geometry and the chemical abundance structures is necessary and the ambiguity due to the presence of element vertical stratification and horizontal inhomogeneities should be kept in mind when evaluating the results published in the literature.

4.3 Rotation

One of the most intriguing properties of the Ap/Bp stars is their slow rotation, compared to the normal main sequence stars in the same temperature range. The rotation periods of these stars can be as short as 0.5 day, but the majority of them have rotation periods in the range of 2–10 days (e.g., Netopil et al. 2017). Some of the Ap/Bp stars have periods that span 5 to 6 orders of magnitude, with no evidence for evolution besides conservation of the angular momentum during their main-sequence lifetime (Mathys 2017).

It is generally assumed that Ap stars are slow rotators because of magnetic braking. Whether they undergo deceleration after arrival on the main sequence (MS) or during the pre-MS phase has been discussed in a number of studies. The evolutionary changes of their rotation periods during their main-sequence lifetimes are small, of the order of a factor of two at most. Thus, period differentiation must have been achieved before the Ap stars reached the main sequence, or even before their progenitors become observable as magnetic Herbig Ae/Be stars. Hartoog (1977) studied the Upper Scorpius region and a number of open clusters, and found no evidence for braking. Therefore, he argued that most of the angular momentum is lost in the pre-MS phase. North (1984) studied silicon-rich stars belonging to clusters and associations as well as to the field and showed that those stars lost most of their angular momentum before the MS phase, or are intrinsically slow rotators from their formation on. No hint of magnetic braking on the MS could be found from a study of the distribution of the rotation periods as a function of log g (North 1985). The studies of Hubrig et al. (2000b, 2007) also did not present any evidence supporting the hypothesis that Ap stars suffer considerable magnetic braking during their MS life. Stępień (2000) suggested that the complex interaction between the stellar magnetic field and the circumstellar environment during the PMS phase could slow down the stellar rotation. In contrast, Hubrig et al. (2000c) investigated a sample of 160 A-type stars with accurate Hipparcos parallaxes and homogeneous determinations of their $v \sin i$-values and concluded that the $v \sin i$ distribution of very young A stars on the main sequence is compatible with a Maxwellian distribution of equatorial velocities. More recently, Netopil et al. (2017) investigated the rotation properties of magnetic Ap stars based on a large sample consisting of 492 objects with known rotation periods. The authors confirmed that the evolution of rotation periods among magnetic stars agrees very well with the assumption of conservation of angular momentum.

Major efforts have been made by Mathys et al. (1997) and Mathys (2017) to identify and study Ap/Bp stars showing spectral lines resolved into their magnetically split components. A by-product of these studies has been the discovery of a sizable population of Ap stars with extraordinary slow rotation. Figure 4.4 from the work of Mathys (2017) shows the distribution of the rotation periods of the Ap stars with resolved magnetically split lines. The shaded part of the histogram corresponds to the 40 stars for which the exact value of the rotation period is known. For the remaining 10 stars, only a lower limit could be determined until now. From consideration of the distribution of the periods of stars with magnetically resolved lines, Mathys (2017) concluded that several percent of all Ap stars must have rotation periods longer than 1 yr, that the periods of some of them must definitely be of the order of 300 yr, and that there may even exist Ap stars with much longer rotation periods, perhaps ~1000yr or more. According to Mathys (2017), the effectiveness of the achievable rotational braking of Ap stars in binaries can also be related to the orbital period (see Section 4.5).

The physical mechanisms responsible for this differentiation of rotation periods, and in particular for the existence of extremely slowly rotating magnetic stars, remain unknown. Obviously, improving the characterisation of these targets is essential to guide and further constrain theoretical developments.

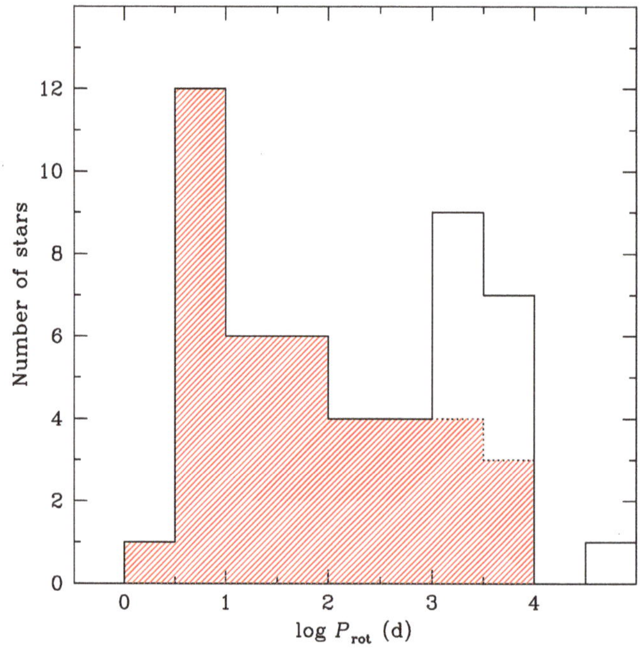

Figure 4.4. Histogram showing the distribution of the rotation periods of the 50 Ap stars with resolved magnetically split lines for which their value has been determined (shaded part of the histogram) or for which a meaningful lower limit has been obtained. Credit: Mathys (2017), reproduced with permission © ESO.

For a few hotter Ap stars with Si overabundance rotation period changes were discussed in several studies. The most detailed monitoring has been carried out for the Ap stars CU Vir and BS Cir. A cyclicity of the period changes or some kind of oscillations of the period around some mean value was proven so far only for CU Vir, with a cycle length of ~66 years and $\frac{d^3P}{dt^3} = 3.25 \times 10^{-16}$ day^{-2} (Mikulášek 2016; Krtička et al. 2019). No interpretation of such changes exists because the mechanism for the observed spin-ups and spin-downs in these stars has not been clarified yet.

4.4 Pulsating Ap Stars

A number of cool Ap stars belong to a subgroup called rapidly oscillating Ap (roAp) stars. The first discovery of a pulsating Ap star was reported by Kurtz (1978), who detected a 12 min pulsation period in Przybylski's star (HD 101065) with $\langle B_z \rangle$ of several hundred Gauss. RoAp stars pulsate in high-overtone, low-degree, non-radial p-modes modes with periods in the range of about 5–25 min. Their pulsational variability is, however, of a very small amplitude, mostly below 0.01 mmag (Kurtz 1982). Compared to the several thousands known Ap stars (e.g., Renson & Manfroid 2009), roAp stars are relatively rare with 61 known to date (Smalley et al. 2015). Out of 61 roAp stars, 14 show spectral lines resolved into their magnetically split components (Mathys 2017). Over the last decades they became extremely promising targets for asteroseismology, which is a most powerful tool for testing theories of stellar structure.

Apart from photometric variability, roAp stars show rapid radial velocity variations with periods similar to those obtained from photometry. The highest pulsation amplitudes of radial velocities were detected for lines of rare earth elements and the Hα core (e.g., Kurtz et al. 2003; and references therein). Spectroscopic observations of the variability in rare-earth spectral lines formed in the outer atmospheric layers of these stars offer unique possibilities for investigating details of the physics of propagating magnetoacoustic waves. The pulsations observed in radial velocity variations of lines of rare earth elements arise at atmospheric heights where the magnetic pressure exceeds the gas pressure and the Alfvén velocity is greater than the acoustic velocity. This means that the observed pulsations are primarily magnetic with an acoustic component. Several studies using times series of high signal-to-noise spectra have allowed to achieve a three-dimensional resolution of the pulsation modes, with the vertical stratification giving depth information, and the rotation of the oblique mode providing information on the surface geometry of the modes (e.g., Kurtz et al. 2004).

Since the discovery of roAp stars, various theoretical studies concentrated on the investigation of mode conversion and reflection, contributions of horizontal and vertical motions to the observed pulsational radial velocities, and mode excitation and suppression (see e.g., Saio 2005; Sousa & Cunha 2011; and references therein). According to numerous theoretical and observational studies, roAp stars are oblique pulsators and the strength and geometry of their global, approximately dipolar magnetic fields constrain the pulsation modes. Their non-radial pulsations have axes that are inclined to the rotation axis of the star and nearly aligned with the magnetic axis (Kurtz 1982; Shibahashi & Takata 1993; Saio & Gautschy 2004; Bigot & Kurtz 2011). On the other hand, Gough (2012) argued that even if the pulsation axis had once been aligned with the magnetic axis, the alignment would be lost because it should precess due to the Coriolis force. To avoid this difficulty, he proposed that pulsation should be excited only when the axis is closely aligned to the magnetic axis.

RoAp stars with very strong magnetic fields are of special interest because the effect of the magnetic field on the structure of their atmospheres can be studied with the greatest detail and accuracy. In particular, the diagnostic potential of time-resolved spectroscopic observations of pulsations is enhanced for roAp stars that have spectral lines resolved into their magnetically split components. Among these stars, the cool roAp star HD 154708 possesses the strongest magnetic field among the roAp stars, with a mean magnetic field modulus $\langle B \rangle = 24.5 \pm 1.0$ kG (Hubrig et al. 2005a, 2006). The magnetic field in this star was discovered during a systematic survey of magnetic fields in Ap/Bp stars with the low-resolution FORS 1 spectropolarimeter installed at the 8-m Kueyen telescope of the VLT (Hubrig et al. 2005a). The presence of pulsations with a period of 8 min was discovered by Kurtz et al. (2006) using high-resolution time series of UVES (Ultraviolet and Visual Echelle Spectrograph) spectra obtained over 2.5 h. The authors reported that the radial velocity amplitudes in the rare earth element lines of Nd II, Nd III, and Pr III were unusually low, suggesting that roAp stars with stronger magnetic fields have lower pulsation amplitudes. With its field strength of $\langle B \rangle = 24.5 \pm 1.0$ kG (Hubrig et al. 2005a), HD 154708 shows by far the most extreme observable case of pulsations in the presence of a strong magnetic field (Kurtz et al. 2006). The magnetic field of HD 154708 is so strong that a number of spectral lines are distorted beyond recognition by the partial Paschen–Back effect. Among the whole class of Ap/Bp stars, the only star known at that time to have a stronger dipolar magnetic field than HD 154708 was the much hotter ($T_{\rm eff} \approx 14,500$ K; Leckrone 1974) Bp Si star HD 215441 (Babcock's star) with a mean field modulus $\langle B \rangle$ of 34 kG (Babcock 1960). Later on, Freyhammer et al. (2008) reported on the presence of a very strong mean magnetic field modulus of 30.3 kG in the non-oscillating Ap star HD 75049 with $T_{\rm eff} = 9600$ K and $\log g = 4.47$ (Elkin et al. 2010), making this the second-strongest magnetic field known among the upper main-sequence chemically peculiar stars.

Thirteen spectropolarimetric observations of HD 154708 were obtained with FORS 1 over three months in 2007. These observations have been used to determine the rotation period $P = 5.3666 \pm 0.0007$ day (Hubrig et al. 2009a). The phase diagram, including the best sinusoidal fit, is shown in Figure 4.5, along with the residuals (lower panel). The size of the error bars is comparable to the size of the dots representing the individual measurements. The scatter of the latter about the best-fit

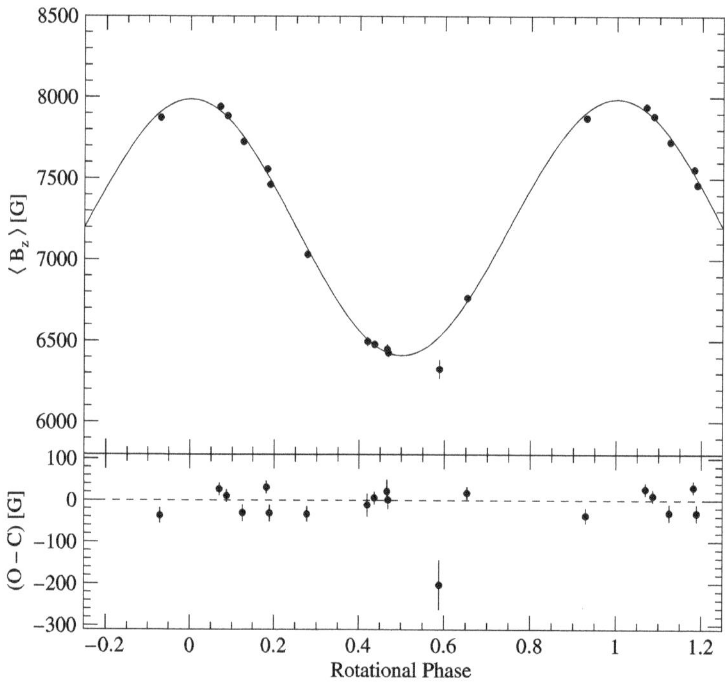

Figure 4.5. Phase diagram for the magnetic field measurements of HD 154708 using the entire FORS 1 spectra for the most accurate period, $P_{\rm rot} = 5.366$ days. The variation has a mean of $\overline{\langle B_z \rangle} = 7199 \pm 11$ G and an amplitude of $A_{\langle B_z \rangle} = 789 \pm 13$ G. The residuals (Observed – Calculated) for the best fit is shown in the lower panel. The deviations are mostly of the same order as the error bars, and no systematic trends are obvious, which justifies a single sinusoid as a fit function. Credit: Figure 3 from Hubrig et al. (2009a).

curve is also of the same order. The measurement with the largest sigma ($\langle B_z \rangle = 6326 \pm 59$ G) was obtained in weather conditions classified as "thick clouds," where the guide star was frequently lost, and thus was repeated by the service observer a couple of nights later. Using stellar fundamental parameters and the longitudinal magnetic field phase curve, it was possible to characterise the magnetic field geometry. The atmospheric parameters of HD 154708, $T_{\rm eff} = 6800$ K and $\log g = 4.11$, have been determined using the B2–G temperature calibration of Geneva photometry (Hauck & North 1993) and the Hipparcos parallax (Hubrig et al. 2005a). On the other hand, Strömgren photometry (Napiwotzki et al. 1993) suggested a hotter temperature, $T_{\rm eff} = 7500$ K. Knowing the position of the star in the H-R diagram (Hubrig et al. 2005a), it was also possible to determine the radius of the star ($R = 1.7 - 2.0\ R_\odot$). According to Hubrig et al. (2009a), HD 154708 is observed nearly pole-on and the magnetic field geometry can be described by a centred dipole with a surface polar magnetic field strength $B_{\rm d}$ between 26.1 and 28.8 kG and an inclination of the magnetic axis to the rotation axis in the range 22.5° to 35.5°.

Interestingly, the comparison of high spectral resolution UVES spectra of HD 154708 at three different dates at a resolving power of $\lambda/\Delta\lambda \approx 1.1 \times 10^5$ available in the ESO archive revealed almost the same appearance of the magnetically resolved Zeeman components. As is shown in Figure 4.6, no noticeable change in the magnetic field modulus between these three observations is detectable. This does not represent a strong constraint: in more than half of the known Ap stars with resolved magnetically split lines, the ratio between the extrema of the mean field

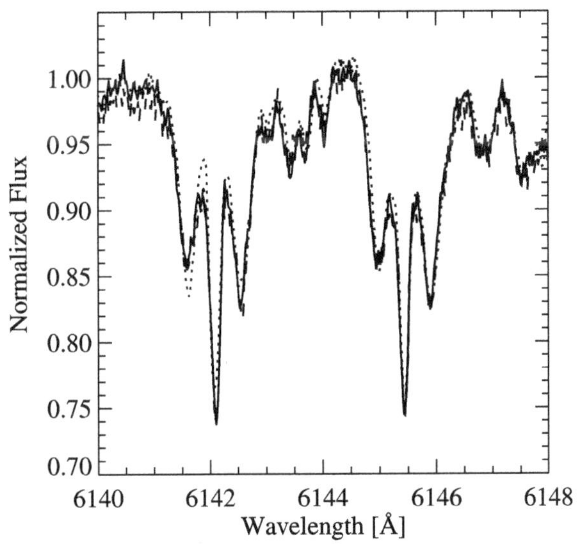

Figure 4.6. No obvious changes are detectable in the split Zeeman structure in the spectra of the roAp star HD 154708 recorded on three different epochs. One spectrum of HD 154708 was obtained on 2005 May 20 at phase 0.72 and is indicated by the solid line. The second spectrum was observed on 2005 September 18 at phase 0.37 and is presented by the dotted line. The third spectrum was observed on 2006 July 23 at phase 0.76 and is presented by the dashed line. Credit: Figure 4 from Hubrig et al. (2009a).

modulus is lower than 1.1, and in 75% of all Ap stars with studied field geometries, it does not exceed 1.3.

The positions in the H-R diagram and the kinematic characteristics of rapidly oscillating and non-oscillating chemically peculiar stars were investigated in detail by Hubrig et al. (2000a) using Hipparcos proper motions and parallaxes, and their own radial velocity measurements. The authors found that rapidly oscillating stars, as a group, are located at −0.47 ± 0.34 mag above the zero-age main sequence (ZAMS), while the non-oscillating stars are at −1.20 ± 0.65 mag above the ZAMS and so appear slightly more evolved on average. From a comparison of the kinematical characteristics, it was concluded that both groups are very similar. Figure 4.7 shows the distribution of roAp and non-oscillating Ap (noAp) stars with available Hipparcos parallaxes in the astrometric H-R diagram (Arenou & Luri 1999), taken from Hubrig et al. (2000b). Obviously, HD 154708 is located in the same region of parameter space in which rapidly oscillating Ap stars have been detected. Further, the results of radial velocity measurements indicated that there was a real deficiency of binaries among rapidly oscillating stars. Presently, no roAp star is known to be a member of a close spectroscopic binary.

To enhance the diagnostics of pulsations in roAp stars, Mathys et al. (2005) studied the behaviour of resolved magnetically split lines in the strongly magnetic roAp star HD 166473. For this star, Mathys & Hubrig (1997) obtained three

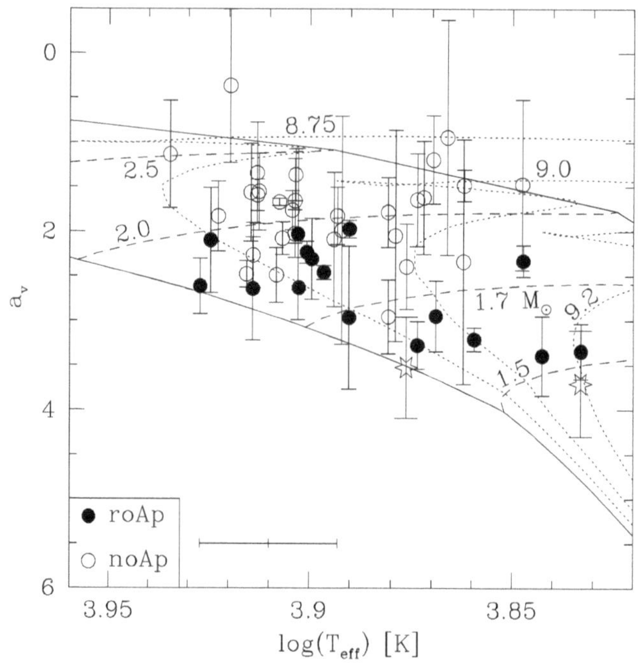

Figure 4.7. Astrometric H-R-diagram of the roAp (filled circles) and of the non-oscillating Ap stars (open circles). HD 154708 is indicated by a 6-branch star at two positions: one (to the left) corresponding to the *uvbyβ* effective temperature of 7500 K, and the other (to the right) corresponding to the Geneva B2–G temperature of 6800 K. Credit: Hubrig et al. (2005a); reproduced with permission © ESO.

measurements of the mean longitudinal field, all of the order of −2 kG. Mathys et al. (1997) reported the discovery of the presence of resolved magnetically split lines and a mean magnetic field modulus $\langle B \rangle$ of 6.4–8.6 kG in HD 166473, whose fundamental parameters, $T_{\rm eff} = 7700$ K and $\log g = 4.2$, were determined by Gelbmann et al. (2000). Kurtz & Martinez (1987) reported the existence of at least three pulsation modes with periods between 8.8 and 9.1 min and rather low photometric amplitudes (less than 0.5 mmag). Mathys et al. (2005) studied for the first time the pulsational behaviour of the individual resolved components using the Eu II λ6437 line and concluded that they possibly detected variations of the mean magnetic field modulus with a frequency of 1.928 mHz and an amplitude of 21 ± 5 G. Unfortunately, the pulsation amplitude in HD 166473 is one of the smallest among roAp stars. Obviously, similar studies of other roAp stars are needed to confirm the prospects for enhancement of the diagnosis of pulsations in roAp stars.

RoAp stars are located in the H-R diagram in and around the δ Sct instability strip. However, their oscillation frequencies are much higher than those of δ Sct variables. The typical pulsation periods of δ Scuti stars range from 20 min to 0.3 day and it is assumed that the κ mechanism acting in the second ionisation zone of helium is responsible for their pulsational behaviour. There is also a number of δ Sct stars that are classified as Ap stars. Pulsations of δ Scuti type among Ap stars are not expected from a theoretical point of view. The models of Saio (2005) led to a clear prediction that lower radial overtone pulsation modes typical in δ Scuti stars are suppressed by the magnetic field in Ap stars and magnetic Ap stars should not be observed to pulsate simultaneously in low radial overtone p modes characteristic for δ Scuti stars and high-overtone p modes characteristic for the roAp stars.

Although pulsations of δ Scuti type are not expected among Ap stars from a theoretical point of view, previous observations of the known δ Scuti star HD 21190 by Koen et al. (2001) indicated a spectral classification F2III SrEuSi. In the course of an analysis of time-series photometry and spectroscopic material of this Hipparcos variable star, the authors discovered pulsations of δ Scuti type, which are unusual among Ap stars. According to the authors, HD 21190 was also one of the most evolved δ Scuti stars known. UVES high time-resolution observations of HD 21190 were used by González et al. (2008) to search for pulsational line profile variations. The spectra of HD 21190 showing broad spectral lines ($v \sin i \approx 72$ km s^{-1}), with small variable peaks in the line profiles are presented in Figure 4.8. In the bottom of this figure all the observed spectra stacked in two-dimensional images for two spectral regions are displayed. González et al. (2008) detected three peaks moving smoothly toward the red with a speed of 21 km s^{-1} per hour. Such a behaviour is typical for non-axisymmetric non-radial pulsations. The combination of the cool Ap spectral type and the high-degree δ Scuti pulsation made HD 21190 an important target for a more in depth study.

Kurtz et al. (2008) used FORS 1 mounted on the 8 m Kueyen telescope of the VLT in spectropolarimetric mode with the GRISM 600B and a slit width of 0.4″ to achieve a spectral resolution of about 2000. The analysis of these data revealed the presence of a rather weak longitudinal magnetic field $\langle B_z \rangle = 47 \pm 13$ G. Hubrig & Schöller (2016) re-observed HD 21190 with FORS 2 in 2016 March, using the same instrumental

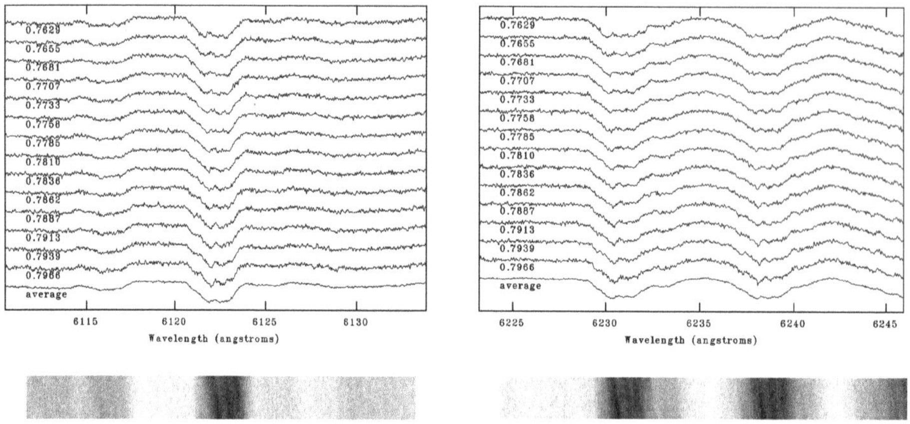

Figure 4.8. Spectral profile variability of the Mn II λ6122 (*left panel*), Fe I λ6231 and Fe II λ6238 (*right panel*) lines in UVES spectra of HD 21190. The spectra are labelled with Julian dates (HJD–2454047). The average spectra are plotted in the bottom of each panel. The two-dimensional images present, for the same spectral regions, the 14 observed spectra stacked. In these images the y-dimension corresponds to the observing time, increasing downwards. Three moving peaks are clearly visible in the cores of all presented spectral lines. Credit: Figure 1 from González et al. (2008).

setup as Kurtz et al. (2008), and detected a magnetic field in HD 21190 at a significance level of more than 4σ ($\langle B_z \rangle_{\rm all} = -254 \pm 59$ G) using for the measurements the entire spectrum, and at a significance level of more than 3σ ($\langle B_z \rangle_{\rm hyd} = -237 \pm 75$ G) using the hydrogen lines. Consecutive FORS 2 magnetic field measurements revealed the presence of a longitudinal magnetic field at a significance level of more than 6σ ($\langle B_z \rangle = 230 \pm 38$ G; Järvinen et al. 2018). These results suggested that the magnetic field of HD 21190 is varying with rotation, as is typical for Ap stars. According to Niemczura et al. (2017) and Järvinen et al. (2018), HD 21190 is currently the only known case of an Ap star with a detected magnetic field that shows δ Scuti pulsations.

Neiner & Lampens (2015) obtained two spectropolarimetric observations of the Kepler δ Sct—γ Dor hybrid candidate HD 188774. One observation showed a weak signal of about 76 G. However, the target did not show any Ap characteristics: the abundance analysis did reveal neither the presence of chemical peculiarities nor of chemical spots. Other discoveries of weak magnetic fields in δ Sct stars were reported by Neiner et al. (2017) and Zwintz et al. (2020), but none of these targets were identified as magnetic Ap stars. Multi-periodic δ Sct pulsations have been detected by Skarka et al. (2019) in the chemically peculiar star HD 99458. δ Sct pulsations have also been detected in the Ap star HD 41641 (Escorza et al. 2016). However, the presence of magnetic fields in these targets was not explored yet.

Although the models of Saio (2005) did not predict the existence of roAp stars showing in addition low-overtone, δ Sct p-mode pulsations, using Kepler observations and spectroscopic observations with the High Dispersion Spectrograph (HDS; Noguchi et al. 2002) on the Subaru telescope, Murphy et al. (2020) discovered the first δ Sct–roAp hybrid pulsator KIC 11296437. The analysis of chemical composition and magnetic field strength measurement showed that this star is an Ap star showing both, high-overtone roAp p modes and low-overtone δ Sct p modes. In this work, the authors also presented new model calculations indicating that magnetic field strengths less than 1 kG do not suppress low-overtone p-modes, and, within the model uncertainties, the fundamental radial mode can be excited even in stars with field strengths of up to 4 kG. It was suggested that other roAp stars with low field strengths should also be found with δ Sct pulsations, similar to KIC 11296437. Further, the calculations showed that low-order g-modes with frequencies similar to the fundamental mode are not suppressed, while the high-order g-modes typical for γ Dor stars are strongly suppressed, explaining the absence of detections of γ Dor pulsations in Ap stars.

4.5 Multiplicity

Among the magnetic Ap stars, the rate of binaries is 43% (Carrier et al. 2002), and the number of SB2 systems is extremely small. The deficiency of short-period binaries compared to normal A-type stars and the lack of circularised orbits was reported by Gerbaldi et al. (1985), Leone & Catanzaro (1999), and Carrier et al. (2002). In the catalogue of Renson & Manfroid (2009) only three Ap bona fide eclipsing system candidates were mentioned: V414 Pup, HY Vir, and V892 Cen. Their orbital parameters do not differ from those of normal stars. The shortest orbital period known for an Ap star is 1.52 days for HD 26961. Only for two double-lined close binary systems with Ap magnetic components, HD 98088 (e.g., Babcock 1958; Abt et al. 1968) and HD 161701 (Hubrig et al. 2014), is the structure of the magnetic field known. Remarkably, in both systems carries the surface of the Ap component, which is facing the companion, a positive magnetic field. It is, however, not clear whether tidal forces may play a role during the dynamic process of tidal synchronisation, to align the field geometries in the observed way. Further systematic searches for magnetic fields in binary systems should be conducted to properly evaluate theoretical models of the origin of their magnetic fields.

Mathys (2017) discovered the existence of a correspondence between the rotation and orbital periods of Ap stars in binaries. Out of the studied 43 Ap stars with magnetically split lines, 22 stars were binaries, indicating a rate of occurrence of binarity in the studied sample of 51%, which is slightly higher than the value of 43% reported by Carrier et al. (2002). Ap components with short rotation periods (i.e., less than about 50 days) were found only in systems with very long orbital periods (more than 1000 days). Conversely, whenever a binary containing an Ap component had a short orbital period (i.e., shorter than 1000 days), the rotation period of the Ap component was long (longer than 50 days). This means that short rotational and orbital periods are mutually excluded. In Figure 4.9 all the studied systems are confined to the upper right region. The dashed diagonal line in this figure, presented by Mathys (2017), is defined by $\log P_{\rm rot} = 3.9 - \log P_{\rm orb}$, which is approximately the lower envelope both in $P_{\rm rot}$ and in $P_{\rm orb}$. Five of the systems have representative points lying above or very near this dividing line. The main exceptions are HD 219749 and the four synchronised systems, HD 5550, HD 25267, HD 25823, and HD 98088, which appear in isolation in the lower left corner of the figure.

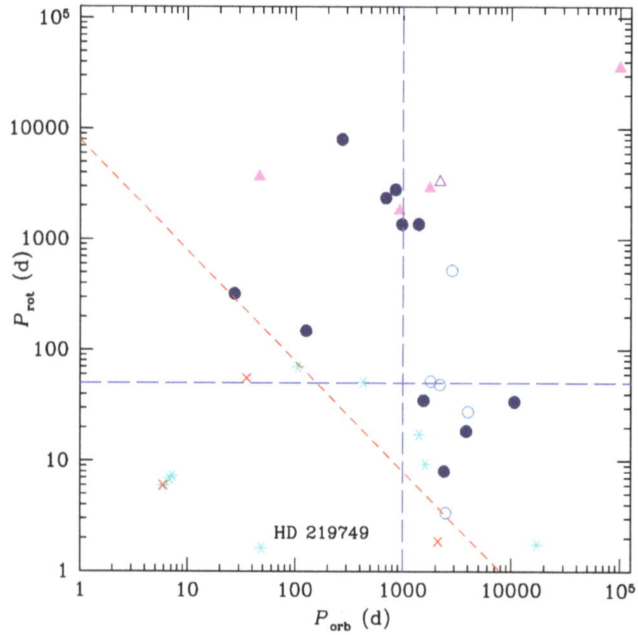

Figure 4.9. Rotation period is plotted against the orbital period for the spectroscopic binaries for which these two periods are sufficiently constrained by existing observations. Asterisks and crosses represent stars not known to have magnetically resolved lines, for which both periods were retrieved from the literature; crosses distinguish those for which uncertainties remain about some of the relevant parameters. All other symbols correspond to Ap stars with magnetically resolved lines studied in detail in Mathys (2017). Dots identify those stars for which both periods are exactly known, open circles stars with fully determined rotation periods for which only a lower limit of the orbital period has been obtained so far, filled triangles stars with a known orbital period but only a lower limit of the rotation period and open triangles stars that have only been observed over a fraction of both their rotation and orbital periods. The short-dashed diagonal line represents the locus $\log P_{rot} = 3.9 - \log P_{orb}$. The long-dashed horizontal line corresponds to $P_{rot} = 50$ days and the long-dashed vertical line to $P_{orb} = 1000$ days. Credit: Mathys (2017), reproduced with permission © ESO.

It is quite possible that the origin of the observed behaviour is related to the earlier evolution of these systems. According to Mathys (2017), the existence of a connection between rotational and orbital periods in Ap binaries represents a very strong argument in support of the merger scenario for the origin of Ap stars (see Section 4.6). Requiring mergers to produce magnetic stars also implies that there should be almost no magnetic star in a close binary. As wide binaries are widespread among Ap stars, such wide binaries could have been born as hierarchical triple stars, where the inner binary merged. A dynamical event as the merger of the two close components of a triple system cannot only determine the rotation rate of the resulting merged star, but also has a significant impact on the orbit of the third component.

Unfortunately, spatially resolved interferometric or adoptive optics observations in the near-infrared of binary systems with magnetic Ap/Bp components are rare although they are extremely useful to characterise the companions to help to understand the role of binarity for the origin of magnetism in stars with radiative envelopes. As an example of such a rare study, the strongly magnetic Bp star HD 94660 with a mass of 3 $M_{\odot}$ and a rotation period of 2800 days (Mathys 2017), was recently observed using the Very Large Telescope Interferometer (VLTI). This star was reported by Bailey et al. (2015) to belong to an SB1 system with an orbital period on the order of 2700 days. Based on the mass function, the secondary in this system was suggested to be rather massive with $M > 2\ M_{\odot}$, yet there was no hint of it in visible spectra. This led Bailey et al. (2015) to the conclusion that the companion might be a compact object, such as a neutron star, a black hole, or a pair of white dwarfs. Schöller et al. (2020) carried out interferometric observations of HD 94660 using PIONIER (Precision Integrated-Optics Near-infrared Imaging ExpeRiment) on the VLTI within the framework of a PIONIER magnetic stars multiplicity survey, aimed at the discovery of wider companions that could have affected the merger event in inner binary pairs. The authors could rule out the hypothesis that the companion to HD 94660 is a compact source and showed that the wide companion is a pair of late-F stars. The presence of lower-mass companions in this system also accounts for the short-time photometric variability detected in TESS data and the X-ray emission detected in Chandra observations, all compatible with a magnetically active companion(s). Future observations of binary stars with Ap/Bp components will represent useful constraints for theoretical modelling of merger processes.

Only one roAp star, HD 42659 has been reported to be a member of a spectroscopic binary with an orbital period of about 93 days (Hartmann & Hatzes 2015; Holdsworth et al. 2019), in direct contrast to the situation for all other cool A stars co-existing in the same region of the H-R diagram, frequently found in SB systems, e.g., δ Sct stars, classical Cepheids (Hubrig et al. 2000a). Interestingly, also no pulsating white dwarf is known to be in an SB system. Neither theoretically nor observationally is present knowledge sufficient to decide confidently whether tidal interaction in binaries may reduce pulsational amplitudes or even inhibit pulsations in cool Ap stars. Tidal forces might conceivably also play a non-negligible role in systems with a larger separation, provided that their eccentricity is large enough. Since tidal forces are strongly dependent on the distance between the components, interaction would then occur mostly on the part of the orbit where the components are closest. On general grounds, the issue of whether duplicity affects pulsation through tidal interaction is unsettled. From the theoretical point of view, while some authors (e.g., Cowling 1941; Zahn 1977) have conjectured that tides in close binary systems may act as an external perturbing force driving oscillations, the question whether tidal interaction may also be efficient in damping already existing pulsations does not seem to have been addressed yet.

A goal of one of the previous programs with NAOS-CONICA (NACO; Lenzen et al. 2003; Rousset et al. 2003) at the VLT in imaging mode from 2007 April to September was to investigate whether roAp stars are indeed single stars or visual binaries. Using the S13 camera, Schöller et al. (2012) discovered 11 visual binaries in the separation range between 0.22 and 10.5$''$ in a sample of 28 roAp stars with K-magnitudes in the range between 4.7 and 10.4. For most systems the separations were found to be rather large, with 6 to 8$''$. The physical nature of the companions is, however, not proven yet. Also almost nothing is known about the orbital eccentricities of the roAp wide binaries. Based on radial velocity measurements, a slightly eccentric ($e = 0.146$) orbit was reported for the roAp star HD 42659 by Hartmann & Hatzes (2015) and Holdsworth et al. (2019). A preliminary visual orbit solution with an eccentricity of 0.56 ± 0.05 was reported for the roAp star HD 201601 by Stelzer et al. (2011).

4.6 Magnetic Field Origin and Evolution of the Field Strength and Geometry

After the discovery of magnetic fields in Ap/Bp stars it was suggested that these stars acquired their field at the time of their formation or early in their evolution and what is currently observed is then a fossil field (e.g., Moss 2003). The competing-dynamo theory proposed that the magnetic field is generated by a turbulent dynamo operating in the star's convective core. As long as it was accepted that strong magnetic fields are observable at all evolutionary states from the ZAMS to the terminal-age main sequence (TAMS), one of the difficulties for the dynamo theory was to explain how the field reaches the stellar surface in the rather short time available before the arrival of the star on the main sequence. Alternatively, magnetic fields may be generated by strong binary interaction, i.e., in stellar mergers, or during a mass transfer or common envelope evolution (e.g., Tout et al. 2008). Two more scenarios for the origin of the magnetic fields of Ap/Bp stars in which these stars result from the merging of two lower mass stars or protostars were suggested by Tutukov & Fedorova (2010) and Ferrario et al. (2009), respectively. The mergers would produce a brief period of strong differential rotation and give rise to large-scale fields in the radiative envelopes. It is therefore quite possible that the binaries with Ap/Bp components that we observe now were triple systems earlier in their history.

One of the fundamental questions of the merger scenario is how the merging binary can lose its angular momentum, i.e., what makes the two stars come closer together. One solution could be a third component that takes over the angular momentum from the inner system and modulates the eccentricity of the inner orbit. In an analysis of the evolution of close triple systems, Mazeh & Shaham (1979) pointed out that when sizable eccentricity-sensitive tidal dissipation exists between a close pair of stars, a third component may bring the pair even closer to each other. Should a wide third component be essential to lead to mergers and thus to Ap/Bp stars with a magnetic field, this wide component should still be physically bound to the magnetic star and be detectable as a wider binary component. Noteworthy, even though stellar mergers are believed to be a common phase in the evolution of binary and multiple systems, observations of such events and details of their evolution remain elusive.

Obviously, understanding the origin and evolution of the magnetic fields of Ap stars, as well as of their slow rotation, requires knowledge of their evolutionary status. Whether the A and B stars become magnetic at a certain evolutionary state before reaching the ZAMS, or during core hydrogen burning, or at the end of their main-sequence life requires systematic studies of field stars with accurate Hipparcos and Gaia parallaxes, established cluster members with known ages, and binary systems.

In a previous study of the evolutionary state of magnetic Ap stars with accurate Hipparcos parallaxes (Perryman et al. 1997) and accurate measurements of the mean magnetic field modulus and mean quadratic magnetic fields, Hubrig et al. (2000b) showed that the distribution of magnetic stars of mass below 3 $M_{\odot}$ differs from that of normal stars in the same temperature range at a high level of significance. Normal A stars occupy the whole width of the main sequence, without a gap, whereas magnetic stars are concentrated toward the centre of the main-sequence band. Further, stars with higher temperature and higher mass in the study showed a trend to possess stronger magnetic fields. A marginal trend of the magnetic flux to be lower in more slowly rotating stars could possibly be seen as suggesting a dynamo origin for the field. The study of Hubrig et al. (2000b) also showed that among stars with strong magnetic fields the strongest fields are found only in the faster rotating stars: all stars with rotation periods exceeding 1000 days exhibit fields below 6.5 kG. A similar result had already been obtained by Mathys et al. (1997) from the consideration of stars with magnetically resolved lines. The most recent study of Mathys (2017) showed that very strong magnetic fields with $\langle B \rangle_{\rm av} > 7.5$ kG occur only in stars with rotation periods of less than 150 days.

Landstreet et al. (2007; and references therein) used FORS 1 to study about 80 Ap/Bp stars that are potential members of open clusters with masses in the range between 2 and 10 $M_{\odot}$. In contrast to the results of Hubrig et al. (2000b, 2007), the authors reported that magnetic fields are present at essentially all evolutionary stages between the ZAMS and the TAMS for all stellar masses between about 2 and 5 $M_{\odot}$. Further, in their study, stars with $M > 3\ M_{\odot}$ showed evidence that the field strength declines by a factor of a few on a timescale of about $2 - 3 \times 10^7$ yr, while stars of $M < 3\ M_{\odot}$ showed no evidence of field strength decrease, nor of magnetic flux decrease, even on a timescale of several times 10^8 yr.

As the whole sample under study of Hubrig et al. (2000b) contained only 33, mostly slowly rotating magnetic stars of mass below 3 $M_{\odot}$, there was a clear need for more magnetic field measurements of faster rotating Ap/Bp stars with accurate Hipparcos parallaxes. To this purpose, Hubrig et al. (2006) used FORS 1 in spectropolarimetric mode for a long-term systematic search for magnetic fields in about 100 upper main sequence chemically peculiar stars with good Hipparcos parallaxes. All studied objects had either Geneva or Strömgren photometry and were chosen in a wider range of masses. Furthermore, the sample included chemically peculiar Ap/Bp stars whose magnetic field had never or only poorly been studied before. Before this study, only 195 Ap stars had reliably measured longitudinal fields, ranging from tens of Gauss to about 20 kG (Bychkov et al. 2003) and only for 114 stars with measured magnetic fields the parallax error was less than 20%.

The distribution of magnetic Ap/Bp stars with accurate parallaxes is shown in Figure 4.10 (Hubrig et al. 2005b). The presented sample includes not only stars with previously known longitudinal fields, but also stars for which magnetic fields were detected using FORS 1 in spectropolarimetric mode. Hubrig et al. (2005b) confirmed

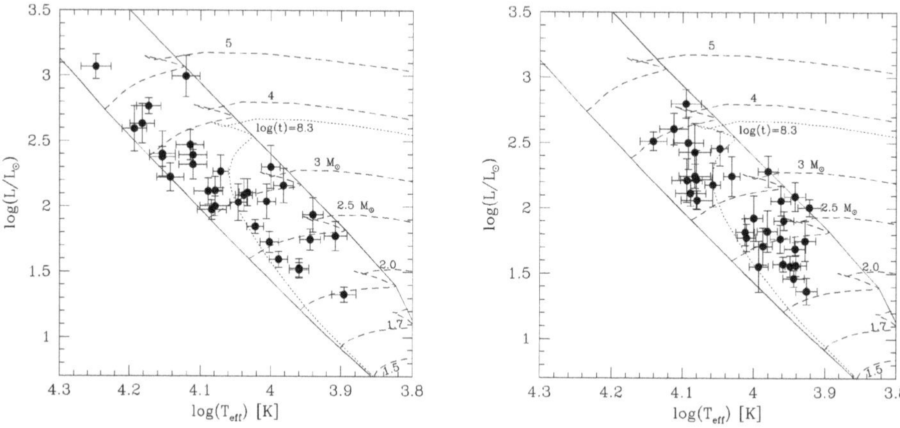

Figure 4.10. *Left panel:* the H-R diagram for the sample of magnetic stars with mean longitudinal fields known from the literature. *Right panel:* the H-R diagram for the sample of magnetic stars with mean longitudinal fields measured using FORS 1. Credit: Figure 1 from Hubrig et al. (2005b).

the previous finding that magnetic stars of mass below 3 $M_{\odot}$ are only rarely found close to the ZAMS, supporting the view that magnetic Ap stars are usually observed in a restricted range of evolutionary states. By contrast, the stars of higher mass seemed to fill the whole width of the main-sequence band. Notably, for Bp stars, the effective temperatures derived from photometry are not in good agreement with the spectral classification (Hubrig et al. 2000b). Because of the extremely anomalous energy distribution and large variability in their spectra, the calibration of the photometric temperature indicators were frequently questioned in the past. Such inconsistencies can only be solved by detailed spectroscopic studies of these stars. The majority of the rotation periods of the studied stars was falling between 2 and 4 days, but there was no indication that the distribution of these faster rotating stars in the H-R diagram differs from that of the slowly rotating magnetic stars studied by Hubrig et al. (2000b).

It is certainly of importance for the considerations of the origin of magnetic fields in Ap/Bp stars that the results of various studies based on very different data sets imply that a certain amount of time is necessary for the magnetic field to build up to become measurable using spectropolarimetry. All the most recently published results on the evolutionary status of Ap/Bp stars, the volume-limited survey of 52 Ap/Bp stars within 100 pc by Sikora et al. (2019b), the survey of 294 Ap/Bp stars by Bernhard et al. (2020) using a colour–magnitude diagram based on the homogeneous Gaia DR2 photometry from Arenou et al. (2018) and the LAMOST DR4 survey by Hümmerich et al. (2020) confirm the concentration of stars with $M < 3M_{\odot}$ toward the centre of the main-sequence band. Bernhard et al. (2020) used archival photometric time-series data from the ASAS-3 (The All Sky Automated Survey), KELT (The Kilodegree Extremely Little Telescope) and MASCARA (Multi-site All-Sky CAmeRA) surveys to improve existing rotation period information and derive rotation periods for Ap/Bp stars hitherto not known to be photometric variables. The recent availability of homogeneous Gaia DR2 photometry from Arenou et al. (2018) and Gaia Collaboration et al. (2018) with more precise G magnitudes and BP–RP indices allowed Bernhard et al. (2020) to place 294 Ap/Bp stars in the colour–magnitude diagram. The distribution of the sample stars in the colour–magnitude-diagram is presented in Figure 4.11 together with PARSEC isochrones (Bressan et al. 2012). Bernhard et al. (2020) emphasised that since their study did not follow the methodological approach of Hubrig et al. (2000b) who used Hipparcos parallaxes and photometric data in Strömgren or in the Geneva system (Hauck & North 1993; Hauck & Kunzli 1996), the observed conspicuous lack of stars directly on the ZAMS cannot be affected by the same potential systematic issues. It should also be kept in mind that any discovery of an Ap star close to the ZAMS does not fundamentally challenge the general results of these studies of the

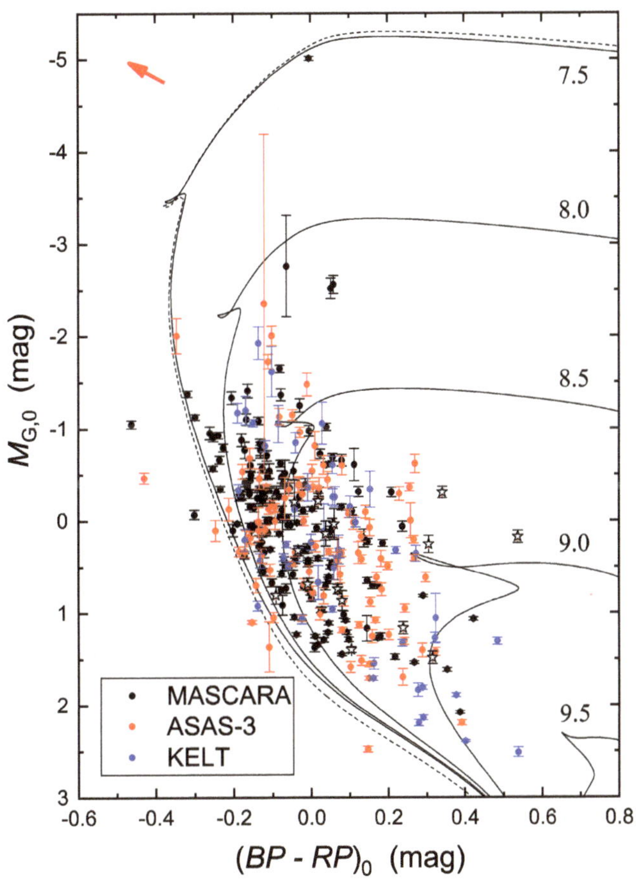

Figure 4.11. The location of the sample stars in the $(BP - RP)_0$ versus $M_{G,0}$ diagram. Also shown are PARSEC isochrones for a solar metallicity of $[Z] = 0.02$ dex (solid lines) and a smaller value of $[Z] = 0.015\,2$ dex (dotted line). Logarithmic ages are given. Open asterisks denote stars for which it was not possible to derive reddening estimates and zero values were adopted. The reddening vector is shown in the upper left corner. Data sources are represented by different colours. Credit: Figure 2 from Bernhard et al. (2020).

evolutionary state of Ap stars since they present statistical studies and do not rule out the existence of some outliers.

The only study of the evolution of magnetic field configurations in a representative sample of Ap/Bp stars was presented by Hubrig et al. (2007). The studied sample consisted of 57 Ap and 33 Bp stars with accurate Hipparcos parallaxes, masses below 6 $M_{\odot}$, known rotation periods and definitely determined periodically variable longitudinal magnetic fields. The sample also included a few known members of nearby open clusters of different ages and with accurate Hipparcos parallaxes. The periods of variation were available in the catalogue of Bychkov et al. (2005) and presented a compilation from the literature and in some cases the determinations of the authors. The distribution of relative ages for the sample of stars with $M < 3\ M_{\odot}$ showed a broad maximum indicating that the majority of these stars have completed 40%–90% of their main-sequence life. The situation was completely different for stars with $M \geqslant 3\ M_{\odot}$: the distribution peaked at the ZAMS age with two secondary lower peaks at relative ages around 60% and 80%. However, given the small number of stars in this sample, these secondary peaks can be considered as only marginally significant and should be confirmed by larger statistics. The cumulative distribution of $\log g$ for both samples is presented in Figure 4.12. A Kolmogorov–Smirnov test implied that the distribution of $\log g$ values for more massive stars differs from the distribution of stars with $M \geqslant 3\ M_{\odot}$ at a significance level of 99.8%. This work also confirmed previous results of various studies that stronger magnetic fields tend to be found in hotter, younger and more massive stars, as well as in stars with shorter rotation periods. No evidence was detected for any loss of angular momentum during the main-sequence life. The magnetic flux was found to be constant over the stellar life time on the main sequence, in agreement with the more recent study of Sikora et al. (2019b).

As for the magnetic field geometry, an excess of stars with large obliquities β (the orientation of the magnetic axis with respect to the rotation axis) was detected in both higher and lower mass stars. The evolution of β appeared to be somewhat different for stars with $M < 3\ M_{\odot}$ compared to stars with $M > 3\ M_{\odot}$, for which a strong hint for an increase of β with time spent on the main sequence was discovered. The geometry of the magnetic fields of the long period Ap stars seems to differ from that of their shorter period counterparts. Landstreet & Mathys (2000) carried out modelling of observed magnetic field moments consisting of the superposition of collinear dipole, quadrupole and octupole for a small sample of slowly rotating magnetic stars with periods $P_{\rm rot} > 25$ days and $M < 3M_{\odot}$. Their result was rather unexpected, implying that the magnetic stars with such periods have small values of obliquity β of the model magnetic axis to the rotation axis, of the order of 20°. The study by Hubrig et al. (2007) of the two stars HD 5737 and HD 168733 with $M > 3\ M_{\odot}$ and the longest periods revealed that they also have magnetic and rotation axes aligned to within 20°, similar to the behaviour of slowly rotating stars of lower mass. Clearly, more magnetic field studies are needed to confirm that the angle between the dipole and the rotation axis tends to be smaller also for slowly rotating stars with higher masses.

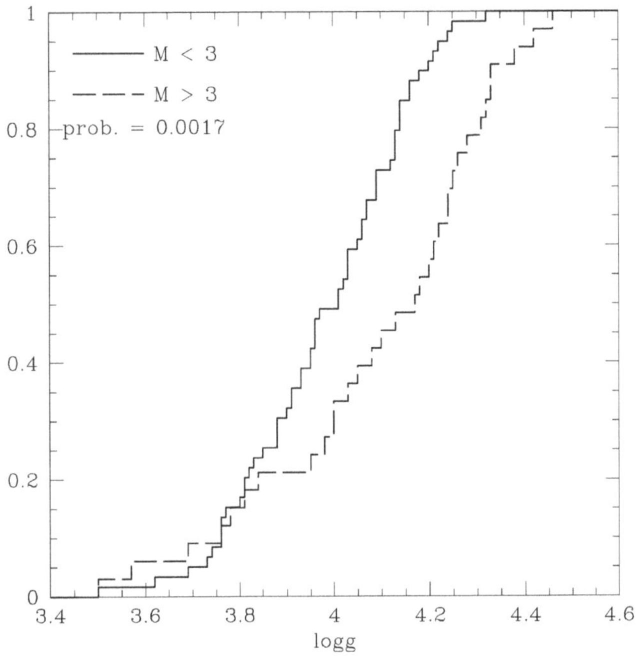

Figure 4.12. Cumulative distribution of log g-values for magnetic stars with masses $M \geqslant 3\ M_\odot$ (dashed line) and $M < 3\ M_\odot$ (solid line). Credit: Figure 3 from Hubrig et al. (2007).

As mentioned by Bychkov et al. (2005), a number of stars in their catalogue exhibit marked anharmonicity, which indicates significant departures from dipole geometry. To reproduce the shapes of the observed longitudinal magnetic field variation, the magnetic field distribution has to be regarded as a sum of dipole and quadrupole components. Out of the 33 stars with $M > 3\ M_\odot$ studied by Hubrig et al. (2007), 21% had magnetic phase curves fitted by a double wave, whereas the percentage of stars with $M < 3\ M_\odot$ for which the magnetic field variations had to be fitted by a double wave was only 12%. This statistics probably indicates that the magnetic topology in higher mass stars is frequently more complex than that in less massive stars. Double-wave longitudinal field curves in stars with $M < 3\ M_\odot$ are mostly found at the age of 40%–50% of their main-sequence life, while a number of stars with $M > 3\ M_\odot$ exhibit a non-dipolar field already at a very young age close to the ZAMS. Further systematic studies of mean longitudinal magnetic fields should be conducted to better assert the evolution of the magnetic field geometry during the main-sequence life.

In summary, although various studies provided new clues, they still remain inconclusive as to the origin of the magnetic fields of Ap/Bp stars. The merging scenario for the origin of Ap stars provides a plausible link between the rotation period of the Ap component of binary systems and their orbital period, but the details of the merging process are expected to be complex. The observation that only 10%–15% of A and B stars are magnetic poses a challenge for both fossil field and dynamo theories. The absence of stars with strong magnetic fields close to the

ZAMS, no observable angular momentum loss before arrival of the stars on the main sequence (Hubrig et al. 2000c), and the absence of strong magnetic fields in Herbig Ae/Be stars (Hubrig et al. 2009b) might be seen as an argument against the fossil field theories. The competing-dynamo theory proposed that the magnetic field is generated by a turbulent dynamo operating in the star's convective core. As long as it was accepted that strong magnetic fields are observable at all evolutionary states from the ZAMS to the TAMS, one of the difficulties for the dynamo theory was to explain how the field reaches the stellar surface in the rather short time available before the arrival of the star on the main sequence. In view of the results presented here that magnetic fields become observable only after the completion of a significant fraction of their main-sequence life, it would be interesting to reconsider the timescales involved in the possible processes of transport of core-generated fields to the surface.

References

Abt, H. A., Conti, P. S., Deutsch, A. J., & Wallerstein, G. 1968, ApJ, 153, 177
Alecian, E., Tkachenko, A., Neiner, C., Folsom, C. P., & Leroy, B. 2016, A&A, 589, A47
Alecian, G., & Stift, M. J. 2017, MNRAS, 468, 1023
Arenou, F., & Luri, X. 1999, in ASP Conf. Ser. 167, Harmonizing Cosmic Distance Scales in a Post-HIPPARCOS Era, ed. D. Egret, & A. Heck (San Francisco, CA: ASP), 13
Arenou, F., Luri, X., Babusiaux, C., et al. 2018, A&A, 616, A17
Aurière, M., Wade, G. A., Silvester, J., et al. 2007, A&A, 475, 1053
Babcock, H. W. 1958, ApJS, 3, 141
Babcock, H. W. 1960, ApJ, 132, 521
Bailey, J. D., Landstreet, J. D., & Bagnulo, S. 2014, A&A, 561, A147
Bailey, J. D., Grunhut, J., & Landstreet, J. D. 2015, A&A, 575, A115
Bernhard, K., Hümmerich, S., Otero, S., & Paunzen, E. 2015a, A&A, 581, A138
Bernhard, K., Hümmerich, S., & Paunzen, E. 2015b, AN, 336, 981
Bernhard, K., Hümmerich, S., & Paunzen, E. 2020, MNRAS, 493, 3293
Bigot, L., & Kurtz, D. W. 2011, A&A, 536, A73
Bressan, A., Marigo, P., Girardi, L., et al. 2012, MNRAS, 427, 127
Bychkov, V. D., Bychkova, L. V., & Madej, J. 2003, A&A, 407, 631
Bychkov, V. D., Bychkova, L. V., & Madej, J. 2005, A&A, 430, 1143
Carrier, F., North, P., Udry, S., & Babel, J. 2002, A&A, 394, 151
Chojnowski, S. D., Hubrig, S., Hasselquist, S., et al. 2019, ApJL, 873, L5
Cowley, C. R., Hubrig, S., Ryabchikova, T. A., et al. 2001, A&A, 367, 939
Cowling, T. G. 1941, MNRAS, 101, 367
Elkin, V. G., Mathys, G., Kurtz, D. W., Hubrig, S., & Freyhammer, L. M. 2010, MNRAS, 402, 1883
Escorza, A., Zwintz, K., Tkachenko, A., et al. 2016, A&A, 588, A71
Ferrario, L., Pringle, J. E., Tout, C. A., & Wickramasinghe, D. T. 2009, MNRAS, 400, L71
Freyhammer, L. M., Elkin, V. G., Kurtz, D. W., Mathys, G., & Martinez, P. 2008, MNRAS, 389, 441
Gaia Collaboration, Babusiaux, C., & van Leeuwen, F. 2018, A&A, 616, A10
Gelbmann, M., Ryabchikova, T., Weiss, W. W., et al. 2000, A&A, 356, 200

Gerbaldi, M., Floquet, M., & Hauck, B. 1985, A&A, 146, 341
González, J. F., Hubrig, S., Kurtz, D. W., Elkin, V., & Savanov, I. 2008, MNRAS, 384, 1140
Gough, D. 2012, GApFD, 106, 429
Hartmann, M., & Hatzes, A. P. 2015, A&A, 582, A84
Hartoog, M. R. 1977, ApJ, 212, 723
Hauck, B., & North, P. 1993, A&A, 269, 403
Hauck, B., & Kunzli, M. 1996, BaltA, 5, 303
Holdsworth, D. L., Saio, H., & Kurtz, D. W. 2019, MNRAS, 489, 4063
Hubrig, S., Kharchenko, N., Mathys, G., & North, P. 2000a, A&A, 355, 1031
Hubrig, S., North, P., & Mathys, G. 2000b, ApJ, 539, 352
Hubrig, S., North, P., & Medici, A. 2000c, A&A, 359, 306
Hubrig, S., Nesvacil, N., Schöller, M., et al. 2005a, A&A, 440, L37
Hubrig, S., North, P., & Szeifert, T. 2005b, in ASP Conf. Ser. 343, Astronomical Polarimetry: Current Status and Future Directions, ed. A. Adamson, C. Aspin, C. Davis, & T. Fujiyoshi (San Francisco, CA: ASP), 374
Hubrig, S., North, P., Schöller, M., & Mathys, G. 2006, AN, 327, 289
Hubrig, S., North, P., & Schöller, M. 2007, AN, 328, 475
Hubrig, S., Mathys, G., Kurtz, D. W., et al. 2009a, MNRAS, 396, 1018
Hubrig, S., Stelzer, B., Schöller, M., et al. 2009b, A&A, 502, 283
Hubrig, S., Carroll, T. A., Gonzalez, J. F., et al. 2014, MNRAS, 440, L6
Hubrig, S., & Schöller, M. 2016, IBVS, 6174, 1
Hubrig, S., Järvinen, S. P., Madej, J., et al. 2018, MNRAS, 477, 3791
Hümmerich, S., Paunzen, E., & Bernhard, K. 2016, AJ, 152, 104
Hümmerich, S., Mikulášek, E., Paunzen, E., et al. 2018, A&A, 619, A98
Hümmerich, S., Paunzen, E., & Bernhard, K. 2020, A&A, 640, A40
Järvinen, S. P., Hubrig, S., Scholz, R. D., et al. 2018, MNRAS, 481, 5163
Järvinen, S. P., Hubrig, S., Mathys, G., et al. 2020, MNRAS, 499, 2734
Khalack, V., & LeBlanc, F. 2015a, AJ, 150, 2
Khalack, V. R., & LeBlanc, F. 2015b, AASP, 5, 3
Kochukhov, O., Bagnulo, S., Wade, G. A., et al. 2004, A&A, 414, 613
Koen, C., Kurtz, D. W., Gray, R. O., et al. 2001, MNRAS, 326, 387
Krtička, J., Mikulášek, Z., Henry, G. W., et al. 2019, A&A, 625, A34
Kurtz, D. W. 1978, IBVS, 1436, 1
Kurtz, D. W. 1982, MNRAS, 200, 807
Kurtz, D. W., Elkin, V. G., & Mathys, G. 2003, MNRAS, 343, L5
Kurtz, D. W., Elkin, V. G., Mathys, G., et al. 2004, in IAU Symp. 224, The A-Star Puzzle, ed. J. Zverko, J. Ziznovsky, S. J. Adelman, & W. W. Weiss (Cambridge: Cambridge Univ. Press), 343
Kurtz, D. W., & Martinez, P. 1987, MNRAS, 226, 187
Kurtz, D. W., Elkin, V. G., Cunha, M. S., et al. 2006, MNRAS, 372, 286
Kurtz, D. W., Hubrig, S., González, J. F., van Wyk, F., & Martinez, P. 2008, MNRAS, 386, 1750
Landstreet, J. D., & Mathys, G. 2000, A&A, 359, 213
Landstreet, J. D., Bagnulo, S., Andretta, V., et al. 2007, A&A, 470, 685
Leckrone, D. S. 1974, ApJ, 190, 319
Lenzen, R., Hartung, M., Brandner, W., et al. 2003, Proc. SPIE, 4841, 944
Leone, F., & Catanzaro, G. 1999, A&A, 343, 273

Luo, A. L., Zhang, H.-T., Zhao, H.-T., et al. 2012, RAA, 12, 1243
Majewski, S. R., Schiavon, S. R., Frinchaboy, P. M., et al. 2017, AJ, 154, 94
Mathys, G. 2002, in Proc. XII Canary Islands Winter School of Astrophysics, Astrophysical Spectropolarimetry, ed. J. Trujillo-Bueno, F. Moreno-Insertis, & F. Sánchez (Cambridge: Cambridge Univ. Press), 101
Mathys, G. 2017, A&A, 601, A14
Mathys, G., & Hubrig, S. 1997, A&AS, 124, 475
Mathys, G., Hubrig, S., Landstreet, J. D., Lanz, T., & Manfroid, J. 1997, A&AS, 123, 353
Mathys, G., & Hubrig, S. 2006, A&A, 453, 699
Mathys, G., Khalack, V., & Landstreet, J. D. 2020, A&A, 636, A6
Mathys, G., Kurtz, D. W., & Elkin, V. G. 2005, in EAS Publications Series, Vol. 17, ed. G. Alecian, O. Richard, & S. Vauclair, 113
Mazeh, T., & Shaham, J. 1979, A&A, 77, 145
Mikulášek, Z. 2016, CoSka, 46, 95
Moss, D. 2003, A&A, 403, 693
Murphy, S. J., Saio, H., Takada-Hidai, M., et al. 2020, MNRAS, 498, 4272
Napiwotzki, R., Schoenberner, D., & Wenske, V. 1993, A&A, 268, 653
Neiner, C., & Lampens, P. 2015, MNRAS, 454, L86
Neiner, C., Wade, G. A., & Sikora, J. 2017, MNRAS, 468, L46
Netopil, M., Paunzen, E., Hümmerich, S., & Bernhard, K. 2017, MNRAS, 468, 2745
Niemczura, E., Scholz, R. D., Hubrig, S., et al. 2017, MNRAS, 470, 3806
Noguchi, K., Aoki, W., Kawanomoto, S., et al. 2002, PASJ, 54, 855
North, P. 1984, A&A, 141, 328
North, P. 1985, A&A, 148, 165
Perryman, M. A. C., Lindegren, L., Kovalevsky, J., et al. 1997, A&A, 500, 501
Preston, G. W. 1969, ApJ, 156, 967
Renson, P., & Manfroid, J. 2009, A&A, 498, 961
Richer, J., Michaud, G., & Turcotte, S. 2000, ApJ, 529, 338
Rousset, G., Lacombe, F., Puget, P., et al. 2003, Proc. SPIE, 4839, 140
Ryabchikova, T., Leone, F., Kochukhov, O., & Bagnulo, S. 2004, in IAU Symp. 224, The A-Star Puzzle, ed. J. Zverko, J. Ziznovsky, S. J. Adelman, & W. W. Weiss (Cambridge: Cambridge Univ. Press), 580
Saio, H. 2005, MNRAS, 360, 1022
Saio, H., & Gautschy, A. 2004, MNRAS, 350, 485
Schöller, M., Correia, S., Hubrig, S., & Kurtz, D. W. 2012, A&A, 545, A38
Schöller, M., Hummel, C. A., Hubrig, S., et al. 2020, A&A, 642, A188
Shibahashi, H., & Takata, M. 1993, PASJ, 45, 617
Shorlin, S. L. S., Wade, G. A., Donati, J. F., et al. 2002, A&A, 392, 637
Sikora, J., David-Uraz, A., Chowdhury, S., et al. 2019a, MNRAS, 487, 4695
Sikora, J., Wade, G. A., Power, J., & Neiner, C. 2019b, MNRAS, 483, 3127
Skarka, M., Kabáth, P., Paunzen, E., et al. 2019, MNRAS, 487, 4230
Smalley, B., Niemczura, E., Murphy, S. J., et al. 2015, MNRAS, 452, 3334
Sousa, J. C., & Cunha, M. S. 2011, MNRAS, 414, 2576
Stelzer, B., Hummel, C. A., Schöller, M., Hubrig, S., & Cowley, C. 2011, A&A, 529, A29
Stępień, K. 2000, A&A, 353, 227

Tout, C. A., Wickramasinghe, D. T., Liebert, J., Ferrario, L., & Pringle, J. E. 2008, MNRAS, 387, 897
Tutukov, A. V., & Fedorova, A. V. 2010, ARep, 54, 156
Wraight, K.-T., Fossati, L., Netopil, M., et al. 2012, MNRAS, 420, 757
Zahn, J. P. 1977, A&A, 500, 121
Zwintz, K., Neiner, C., Kochukhov, O., et al. 2020, A&A, 643, A110

Swetlana Hubrig and Markus Schöller

Chapter 5

O- and Early B-type Stars

Massive stars drive energetic physical processes that affect the structure of entire galaxies and chemically enrich the ISM. They are important sources of ionising photons and elements heavier than helium and determined the early evolution of the universe. It was shown in the past decades that massive stars possess large-scale organised magnetic fields and that these fields can be as weak as a few Gauss. Most theoretical work in the last decades concentrated on the significance of the effects of magnetic braking and magnetic wind confinement.

5.1 General Background

Massive stars with initial masses ranging from 8 to 100 $M_{\odot}$ are of particular interest as they end their evolution with a supernova explosion, producing neutron stars or black holes. Magnetic O stars with masses larger than 30 $M_{\odot}$ and their WR descendants have been suggested as progenitors of magnetars (e.g., Gaensler et al. 2005). A magnetic mechanism for the collimated explosion of massive stars relevant for long-duration gamma-ray bursts (GRBs), X-ray flashes (XRFs) and asymmetric core collapse supernovae was proposed by Uzdensky & MacFadyen (2006). Also observations of extremely bright and energetic supernovae suggest an important role of magnetic fields in the explosion mechanism of a fraction of massive stars (Smith et al. 2011; Nomoto et al. 2011). Magnetised winds can spin down a star by applying a torque at the surface (ud-Doula et al. 2008; Meynet et al. 2011) and magnetic fields can also mediate angular momentum transport in the stellar interior (Heger et al. 2005; Maeder & Meynet 2005). A better understanding of these processes will provide strong constraints on the role of rotationally induced mixing in massive stars (Brott et al. 2011) and on the frequency of long-duration gamma-ray bursts (Yoon et al. 2006).

The past two decades have seen a significant step forward in our understanding of the occurrence of magnetic fields in massive stars. This became possible through the MAGORI (MAGnetic field ORIgin; Hubrig et al. 2011e, 2013), MiMeS

doi:10.1088/2514-3433/abefccch5

(The Magnetism in Massive Stars; Wade et al. 2011) and BOB (B-fields in OB stars; Morel et al. 2015; Hubrig et al. 2014b) surveys, which made use of the currently available spectropolarimeters ESPaDOnS at the Canada–France–Hawaii telescope, NARVAL at the Observatory of Pic-du-Midi, FORS 2 at the ESO/VLT and HARPSpol at ESO/La Silla. While the MiMeS survey was based on high-resolution spectropolarimetry with ESPaDOnS, NARVAL, and HARPSpol, the MAGORI survey used low-resolution FORS 2 spectropolarimetry. The BOB survey used both, FORS 2 and HARPSpol spectropolarimetry. These surveys indicated that probably about 7% of O-type stars with masses exceeding 18 $M_{\odot}$ and about 6% of early-B and O stars have measurable magnetic fields, with field strengths in the range from a few tens of Gauss to a few tens of kG (e.g., Grunhut et al. 2017; Schöller et al. 2017). The majority of the magnetic massive stars exhibits a smooth, single-wave variation of the longitudinal magnetic field during the stellar rotation cycle. These observations are considered as evidence for a dominant dipolar contribution to the magnetic field topology.

The fraction of magnetic stars is probably higher than reported in the studies of Grunhut et al. (2017) and Schöller et al. (2017). Braithwaite & Spruit (2004) showed that stars with radiative envelopes can host stable large-scale magnetic field configurations, as those observed in the Ap/Bp stars, independently of the star's rotation. The implication of their study is that up to about 30%–50% of all massive stars might be magnetic. So far, only the tip of the iceberg has been seen (Stello et al. 2016). In line with these results, the study of the BOB consortium aiming to establish a lower boundary of magnetic fields in massive stars revealed the presence of very weak Zeeman signatures across the line profiles in high-quality HARPSpol observations of two early B-type stars (Fossati et al. 2015). This work indicated that also weak magnetic fields of the order of a few Gauss can be present in massive stars, but possibly escape their detection due to insufficient measurement accuracy.

5.2 Magnetic O Stars

While several tens of early B-type stars were detected as magnetic already several decades back, the first magnetic field detection in an O-type star was achieved only two decades ago, even though the existence of magnetic O-type stars had been suspected for a long time. Indirect observational evidence for the presence of magnetic fields were the many unexplained phenomena observed in massive stars that were thought to be related to magnetic fields, such as cyclical wind variability, Hα emission variation, chemical peculiarity, narrow X-ray emission lines, and non-thermal radio/X-ray emission. However, direct measurements of the magnetic field strength in massive stars using spectropolarimetry to determine the Zeeman splitting of spectral lines are difficult as only a rather small number of spectral lines are available for the measurements. Additionally, spectral lines in massive stars are usually strongly broadened by rapid rotation and macroturbulence, and frequently appear in emission or display P Cygni profiles. Furthermore, apart from the Potsdam Echelle Polarimetric and Spectroscopic Instrument (PEPSI) attached to the Large Binocular Telescope (LBT) in Arizona and designed to utilise both 8.4 m apertures, all other existing high-resolution spectropolarimeters are operating at telescopes with smaller apertures and cannot deliver the necessary high signal-to-noise (S/N) observations for a majority of the massive stars.

As of today, about a dozen O stars are known to possess oblique, predominantly dipolar magnetic fields with surface magnetic field strengths in the range between a few tens of Gauss to more than ten kG. The first detection of a magnetic field in the massive O6pe star θ^1 Ori C (HD 37022) was announced by Donati et al. (2002). The rotation period of 15.426 days of this star based on the variability of spectral lines was determined by Stahl (1998). The variability of the mean longitudinal magnetic field over the rotation period was reproduced by a 1.1 kG dipole field inclined at 42° to the rotation axis (Donati et al. 2002). The Stokes I and V profiles obtained by Donati et al. (2002) applying the technique of least-squares deconvolution (LSD) are presented in Figure 5.1. It shows the strongest and weakest magnetic signals recorded, respectively, on 1999 December 20 and 29, i.e., on rotational phases 0.033 and 0.621.

Four years later Donati et al. (2006) detected a magnetic field with a polar strength of about −1.5 kG in the very slowly rotating ($P_{\rm rot}$ = 538 days) spectroscopically peculiar O6.5f?p star HD 191612, which is a primary in a double-lined

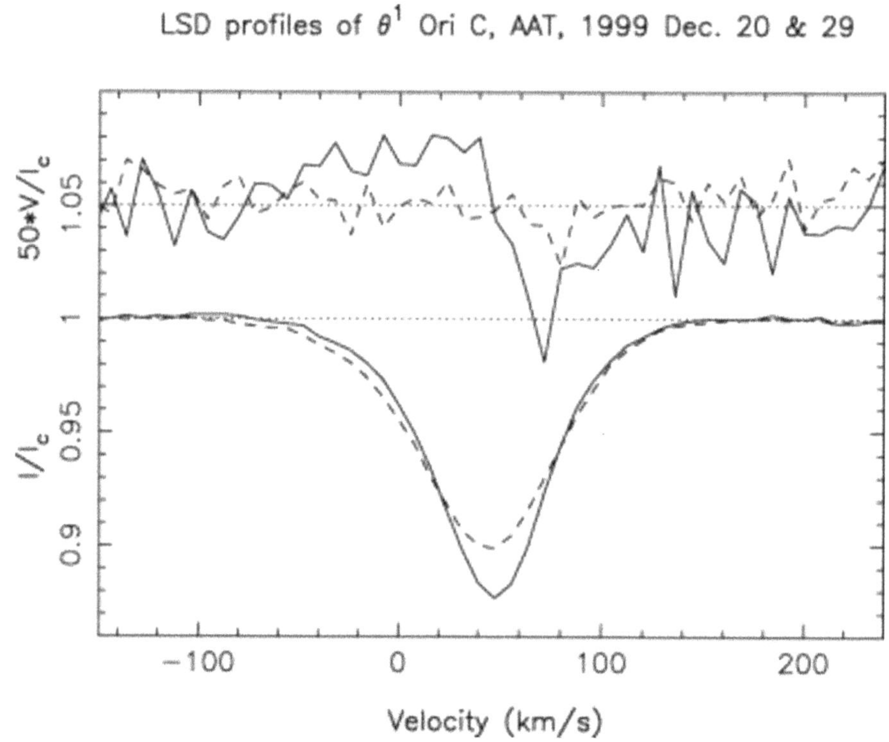

Figure 5.1. LSD Stokes I (*bottom curves*) and Stokes V (*top curves*) profiles of θ^1 Ori C at the rotation phase 0.033 (full line) and 0.621 (dashed line). Note that Stokes V profiles were expanded by a factor of 50 and shifted upwards by 1.05 for clarity. Credit: Figure 1 from Donati et al. (2002).

spectroscopic binary with an orbital period of 1542 days. The peculiar characteristics of the rare class of Of?p stars were discussed by Walborn (1972, 1973), who was also the first to suggest that stars with this category possibly possess magnetic fields. Of?p stars are characterised by recurrent, and apparently periodic, spectral variations in Balmer, He I, C III, and Si III lines, sharp emissions or P Cygni profiles in He I and the Balmer lines, and a strong C III blend in emission around 4650 Å (Walborn 1972). The presence of variable emission in the C III blend and in Hα in Of?p stars is indicative of a circumstellar structure related to their magnetospheres.

Only five Of?p stars are known in our Galaxy and it was shown in the last 15 years that all Of?p stars are magnetic with field strengths from a few hundred Gauss to more than 10 kG. Following the first discovery of a magnetic field in the O6.5f?p star HD 191612, magnetic fields were successively detected in the other four Of?p stars. A subsequent detection of the magnetic nature of the second O6f?p star, HD 148937, was achieved by Hubrig et al. (2008) using the low-resolution VLT instrument FORS 1. In view of the large line broadening in massive stars, to search for the presence of magnetic fields, the low-resolution VLT instruments FORS 1 and FORS 2 (the FOcal Reducer low-dispersion Spectrographs; Appenzeller et al. 1998) at the VLT are considered to be especially suitable instruments, offering the appropriate spectral resolution and the required spectropolarimetric sensitivity, giving access to massive stars even in galaxies in our neighbourhood. The spectropolarimetric capabilities of FORS 1 were moved to FORS 2 in 2009. The currently available low-resolution FORS 1/2 (Hubrig et al. 2008, 2011e, 2013) and high-resolution ESPaDOnS observations (Wade et al. 2012b) of HD 148937 are presented in Figure 5.2. This figure demonstrates the excellent agreement between the FORS 1/2 and ESPaDOnS measurements, highlighting the outstanding potential of FORS 1/2 for the detection of magnetic fields and the investigation of the magnetic field geometry in massive stars. Notably, while an exposure time of 21.5 h at the CFHT with ESPaDOnS was necessary to obtain seven binned measurements, the exposure time for the individual FORS 1/2 observations usually accounted only for two to four minutes and only 2.3 h were used for the observations at six different epochs, including telescope presets and overheads for readout time and retarder waveplate rotation.

Among the five Of?p stars, only two stars, HD 148937 and CPD−28 2561 are observable from the Southern hemisphere. Noteworthy, the first magnetic field detections in these stars were achieved through FORS 1 and FORS 2 observations (Hubrig et al. 2008, 2011e). Similar to the variability of the mean longitudinal magnetic field in HD 148937 presented in Figure 5.2, the FORS 2 measurements of the O6.5f?p star CPD−28 2561 were consistent with a single-wave variation over the stellar rotation cycle, indicating a dominant dipolar contribution to the magnetic field topology. The estimated polar strength of the surface dipole accounted for $B_{\rm d} \geqslant 1.15$ kG (Hubrig et al. 2011e, 2015b). Importantly, in the FORS 1/2 observations of these two Of?p stars, none of the reported detections reached a 4σ significance level. While low-σ detections with low-resolution instruments cannot always be trusted for single observations, they are genuine if the measurements show

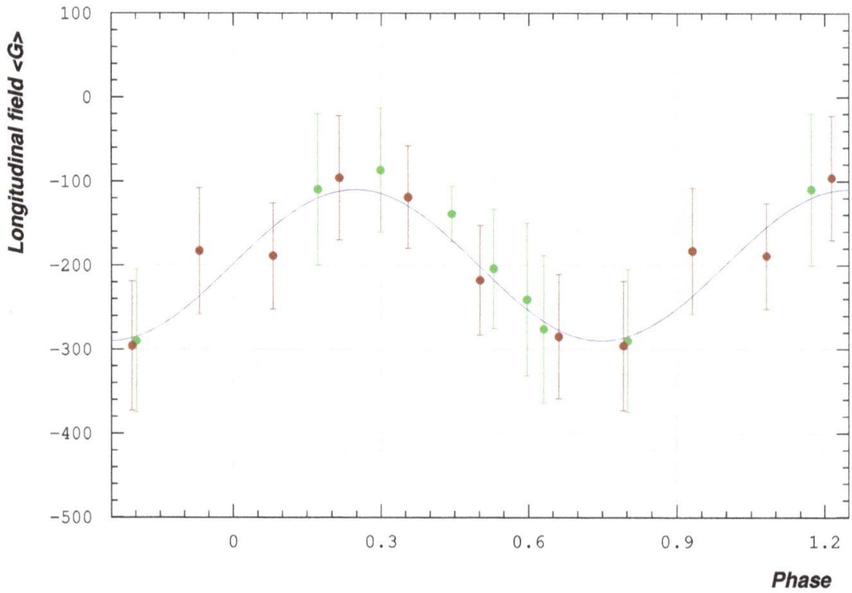

Figure 5.2. Longitudinal magnetic field variation of the Of?p star HD 148937 according to the 7.032 days period determined by Nazé et al. (2008). Red symbols correspond to ESPaDOnS observations (Wade et al. 2012a), while green symbols are FORS 1 and FORS 2 measurements (Hubrig et al. 2008, 2011e, 2013). Note that the measurement errors for both ESPaDOnS and FORS 1/2 observations are of similar order. Credit: Hubrig et al. (2013b), reproduced with permission © ESO.

smooth variations over the rotation period, similar to those found for HD 148937 and CPD−28 2561.

The detections of a magnetic field in the remaining two Of?p stars, the O4f?p star HD 108 and the O7f?p star NGC 1624-2, using high-resolution ESPaDOnS and Narval observations were reported by Martins et al. (2010) and by Wade et al. (2012b), respectively. The longitudinal magnetic field of the extremely slowly rotating star HD 108 with a period of at least a few decades and a polar large-scale magnetic field of about 0.5 kG was found to increase slowly from 100 to 150 G over three years. The slowly rotating ($P_{\rm rot}$ = 158 days) O7f?p star NGC 1624-2, is of special interest as it possesses the strongest magnetic field detected among O stars, and a giant magnetosphere. Its longitudinal magnetic field of positive polarity of the order of several kG was detected by Wade et al. (2012b). Due to the presence of such a strong magnetic field and the slow rotation, a few spectral lines show magnetic splitting into Zeeman components, permitting to investigate the variability of the mean magnetic field modulus over the rotation period (e.g., Järvinen et al. 2021). Obviously, any additional information on the properties of NGC 1624-2 is of particular importance as it is the only magnetic O star for which the variability of both the longitudinal magnetic field and the mean magnetic field modulus can be studied over the rotation cycle.

The results of the most recent magnetic field measurements obtained applying the LSD technique to ESPaDOnS and Narval spectra and using different line samples of photospheric absorption lines, including the Si lines, a line sample without the Si lines and measurements using the high-excitation emission line O III λ7455 are presented in Figure 5.3 (Järvinen et al. 2021). The measured $\langle B_z \rangle$ values are in the range from −0.2 to 4.5 kG. In contrast, as shown in Figure 5.4, the measured mean longitudinal magnetic field using the emission lines He I λ5876 and He I λ6678 for the most part corresponds to the negative polarity. Assuming a dipolar magnetic field structure, the estimated dipole strength of NGC 1624-2 is $B_d \geqslant 16$ kG (Järvinen et al. 2021).

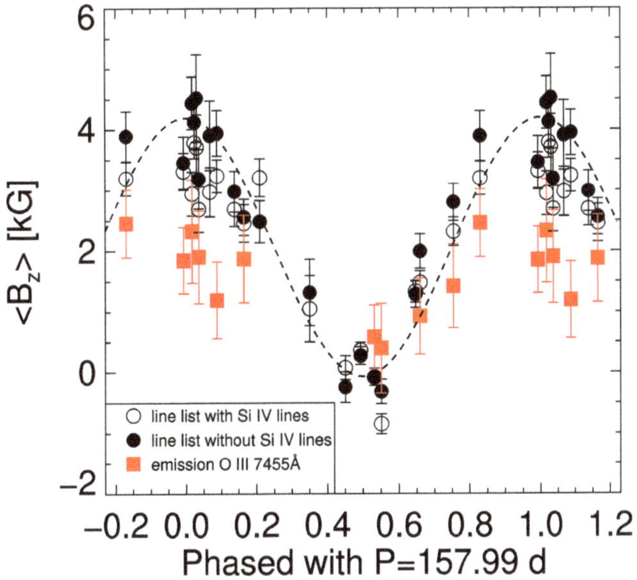

Figure 5.3. Mean longitudinal magnetic field $\langle B_z \rangle$ of NGC 1624-2 measured using two different line samples and the high-excitation line O III λ7455 against rotation phase. The dashed line represents the best sinusoidal least-squares fit solution for the phase curve defined by the magnetic field measurements. Credit: Figure 1 from Järvinen et al. (2021).

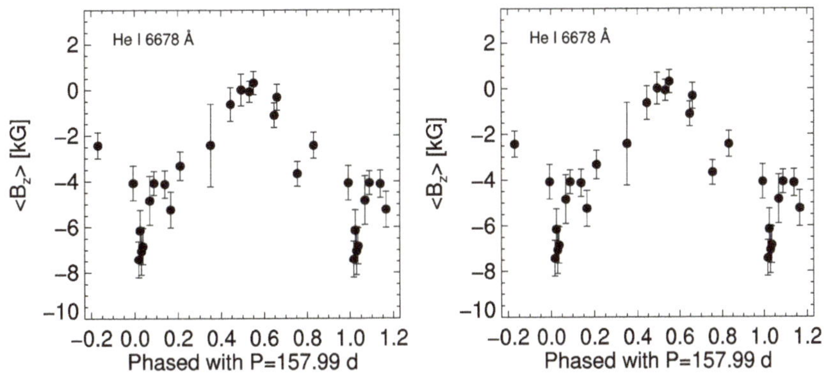

Figure 5.4. As Figure 5.3, but $\langle B_z \rangle$ measured using the He I λ5876 and He I λ6678 lines. Credit: Figure 4 from Järvinen et al. (2021).

Similar to other Of?p stars, line profiles in the spectra of NGC 1624-2 show variability with the same period as the mean longitudinal field. The dynamic spectra of emission and photospheric lines show a single-wave variation of their intensities. The intensity of the lines Hα, He II λ6486, He I λ5876 and He I λ6678 are the strongest close to the positive extremum of $\langle B_z \rangle$ whereas the photospheric absorption lines belonging to He I, C IV, Si IV, and the emission O III lines are the strongest close to the rotation phase 0.5. Examples of the variability of different lines are presented in Figure 5.5.

Measurements of the mean magnetic field modulus $\langle B \rangle$, which is the average over the visible stellar hemisphere of the modulus of the magnetic field vector, weighted by the local line intensity, are presented in Figure 5.6. The displayed results of the measurements are based on ESPaDOnS, NARVAL, and PEPSI polarimetric spectra. The field modulus was calculated by fitting the magnetically resolved C IV λ5812 line with two Gaussian profiles (see Chapter 2). A comparison of the variation curves $\langle B_z \rangle$ and $\langle B \rangle$ indicates that the minimum of the mean magnetic field modulus corresponds to the maximum of the mean longitudinal magnetic field, but slightly shifted in phase with respect to the positive extremum of $\langle B_z \rangle$. This behaviour points to the geometrical structure of the magnetic field of NGC 1624-2 departing from a single dipole at the centre of the star and requires contributions of additional multipole components. A future confirmation of the presence of a shift is necessary, as its presence would hint at a non-axisymmetric magnetic field configuration in this star.

The presence of magnetic fields was also detected in stars of other classification categories. A very weak longitudinal magnetic field with a surface dipolar field strength of about 60 G was detected in the O9.7 supergiant ζ Ori A (Bouret et al. 2008). Grunhut et al. (2009) measured a longitudinal magnetic field of the order a few hundred Gauss and a tentatively derived dipole field strength of 1.7 kG in the O9 sub-giant star HD 57682. Karitskaya et al. (2009) reported the presence of a longitudinal magnetic field of the order of up to 130 G in the optical component with spectral type O9.7 in the system Cyg X-1 using FORS 1 observations (see for more details Section 7.7.1). The detection of a strong ($B_d \sim 2.85$ kG) large-scale organised magnetic field in the secondary component of the massive O8III/I+O7.5V/III

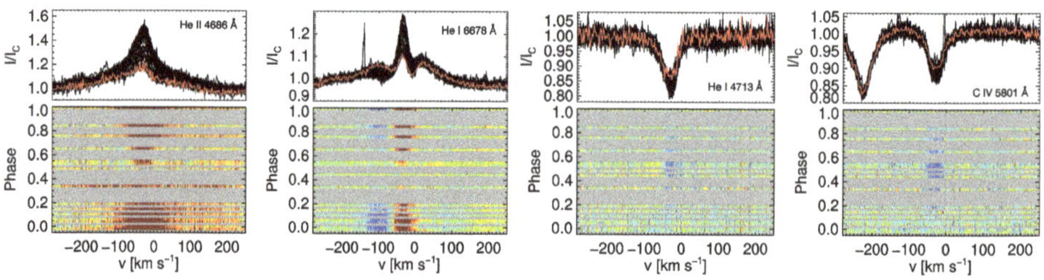

Figure 5.5. Examples of line intensity variability of the emission lines He II λ4686 and He I λ6678 and of the photospheric absorption lines He I λ4713 and C IV λ5801 as a function of rotation phase for NGC 1624-2. All profiles overplotted with the observed weakest profile highlighted in red are presented in the upper panel. Credit: Figure 2 from Järvinen et al. (2021).

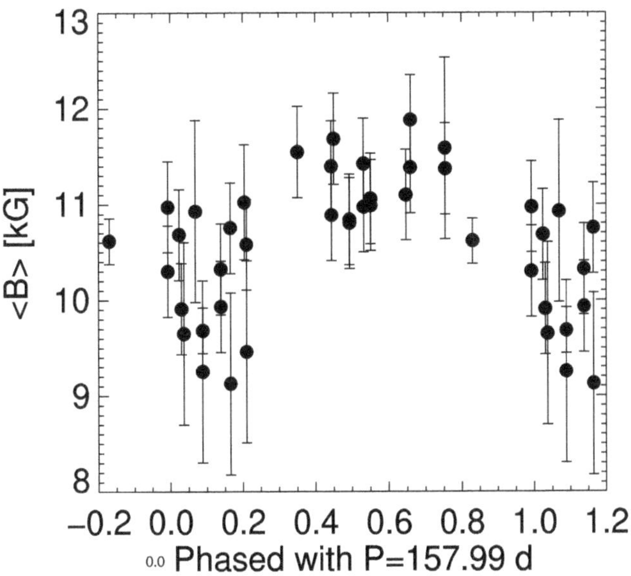

Figure 5.6. Mean magnetic field modulus $\langle B \rangle$ of NGC 1624-2 measured using the split components of the magnetically resolved line C IV λ4812 against the rotation phase. Credit: Figure 2 from Järvinen et al. (2021).

colliding-wind double-lined spectroscopic binary system HD 47129 (Plaskett's star) was reported by Grunhut et al. (2013). Using FORS 2 low-resolution spectropolarimetry, a mean longitudinal magnetic field of the order of a few hundred Gauss was detected in Tr 16-22 by Nazé et al. (2012). The field detection for this star was already expected, as its X-ray emission is bright, variable and hard, compared to other single O-type stars or systems with O-type companions. The most recent detection of a magnetic field in an O-type star, the O9.7V star HD 54879 with a dipole strength larger than 2 kG, was reported by Castro et al. (2015) using a combination of low-resolution FORS 2 and high-resolution HARPSpol observations.

Generally, magnetic O stars exhibit a smooth, single-wave variation of the longitudinal magnetic field during the stellar rotation cycle. The approximately sinusoidal variation of $\langle B_z \rangle$ and the ratio of the values of the $\langle B \rangle$ extrema suggest that there is an important component of the field that is dipolar. Several magnetic O-type stars show the absence of a polarity sign reversal of the mean longitudinal field over the rotation cycle, indicating that only one magnetic pole is visible while the star rotates. Among the magnetic O-type stars with known $\langle B_z \rangle$ variation curves, for seven stars, HD 108, HD 37022 (= θ^1 Ori C), HD 54879, HD 148937, HD 191612, Tr 16-22, and NGC 1624-2, only one magnetic pole is visible throughout the rotation cycle (Shultz & Wade 2017; Petit et al. 2008; Hubrig et al. 2020; Wade et al. 2012a; Donati et al. 2006; Nazé et al. 2016, Järvinen et al. 2021). This implies that a larger fraction of their surfaces can never be observed, leaving the structure of the field over the invisible surface only constrained by the assumption of a tilted centred dipole. There will thus always be a considerable degree of ambiguity left in the magnetic models of these stars.

The majority of the magnetic O-type stars have significantly longer rotation periods compared to those estimated for non-magnetic O stars (e.g., Grunhut et al. 2017), indicating the decisive role of magnetic fields for the loss of angular momentum. According to ud-Doula et al. (2009), the magnetic field strengths required to spin down a massive star are of the order of a few hundred Gauss. Magnetic massive stars also frequently show a very low macroturbulent velocity, which can be understood as a consequence of a magnetic field strong enough to stabilise the atmosphere at deep sub-photospheric layers, where the iron-opacity peak is reached and macroturbulence is thought to originate (e.g., Cantiello et al. 2009).

5.2.1 Small-scale Magnetic Fields

A number of theoretical and observational studies suggest that small-scale magnetic fields can also exist on the surfaces of massive stars (Cantiello & Braithwaite 2011; Braithwaite & Cantiello 2013). Previous theoretical stellar structure calculations of hot massive stars had suggested the presence of subsurface convectively unstable regions related to the iron-opacity peak (Cantiello et al. 2009). Such subsurface convection zones can generate localised magnetic fields associated with brightness spots visible at the stellar surface (Cantiello & Braithwaite 2011). Indeed, spectroscopic studies of massive stars reveal a variety of spectral variability: discrete absorption components (DACs) in ultraviolet (UV) P Cygni profiles, optical line-profile variability, radial velocity modulations, etc. Spectroscopic time series observations frequently show different absorption features propagating with different accelerations. Such features are thought to be signatures of corotating interaction regions (CIRs) in the winds (Mullan 1984). Although DACs are ubiquitous among O-type stars, only for a number of massive stars, among them ξ Per, HD 34078, HD 34656, HD 64760, HD 150168, 68 Cyg, ζ Pup, and λ Cep, the analysis of dynamic UV spectra secures that the variability has a photospheric origin and does not originate from above the stellar surface (e.g., Massa & Prinja 2015). While DACs are understood to be induced by irregularities on the stellar surface, related either to temperature spots—magnetic loops probably due to localised magnetic activity, more or less in analogy to what is observed on the solar surface—or non-radial pulsations (Cranmer & Owocki 1996), CIRs are spiral-shaped density and velocity perturbations in the stellar wind that can extend outward up to several tens of stellar radii. Cranmer & Owocki (1996) carried out two-dimensional hydrodynamical simulations to model the wind response to large-scale base perturbations. They found a structure induced by a localised increase in the radiative flux resulting in bright star spots.

For stars showing CIRs, it has been known for a long time that the absorption features seen in dynamic spectra must be huge in order to remain in the line of sight for days. This means that the linear size of the region responsible for the excess absorption must be at least 15%–20% of the stellar diameter and, in some cases, considerably larger. Ramiaramanantsoa et al. (2014) used the MOST microsatellite to obtain four weeks of continuous high-precision broadband visual photometry of ξ Per, which has the most extensive observational record regarding wind and atmosphere connection, DAC information, and pulsation details. This star also exhibits short-period line-profile variations, implying the presence of non-linear pulsations with a period of 3.5 h. From the analysis of light variations based upon a spot model, the authors deduced the presence of a few corotating (presumably magnetic) bright spots on the stellar surface with a period of 4.18 days, the best-estimated rotation period for this star. The presence of spots is also consistent with theoretical predictions that the lifetime of the magnetic field that could be generated in the subsurface convective zone of massive stars is the turnover time of this convective layer (Cantiello & Braithwaite 2011), which for ξ Per is of the same order as the rotation period. According to Cantiello & Braithwaite (2011), a subsurface convection zone due to partial ionisation may be the source of a global magnetic field, winding up toroidally with stochastic buoyancy breakouts at the surface, causing corotating magnetic bright spots at the surface of the star.

Because the case of ξ Per is not unique regarding DACs and CIRs, magnetic bright spots as indicators of a global magnetic field should be observable in other O-type stars showing similar characteristics and for some stars (e.g., HD 34078, λ Ori A, and ζ Pup) a photospheric origin of their variability was suggested. ζ Pup is a bright, rapidly rotating, runaway supergiant and due to its brightness was frequently monitored in a wide range of wavelengths. A possible presence of a photometric period of 1.78 days in Coriolis/SMEI data was suggested by Howarth et al. (2014) and was later confirmed by BRITE photometry (Ramiaramanantsoa et al. 2018). Recent Chandra X-ray observations of ζ Pup with a long exposure time also showed clear variability with a period previously reported in optical photometric studies (Nichols et al. 2021).

As a possible candidate for the variability detected in multiwavelength observations of ζ Pup, Ramiaramanantsoa et al. (2018) suggested the surface emergence of small-scale magnetic fields generated through dynamo action within a subsurface convection zone associated with the iron-opacity bump at $T \approx 150\,000$ K. Spectropolarimetric monitoring of ζ Pup using low-resolution FORS 2 spectropolarimetry was performed by Hubrig et al. (2016), who found modulation of the longitudinal magnetic field measurements when they phase-folded their data with the 1.78 days period. The mean longitudinal magnetic field variability of ζ Pup over a period of 1.78 days is presented in Figure 5.7, for absorption and emission lines combined, absorption lines alone, and for hydrogen lines. The solid line represents a fit to the joined three sets of measurements, with a mean value of the magnetic field of $\overline{\langle B_z \rangle} = 28.2 \pm 10.5$ G, an amplitude of $A_{\langle B_z \rangle} = 104.0 \pm 19.2$ G and a root mean square value of 62.2 G. For the presented fit, a zero phase corresponding to the negative field extremum at MJD56 570.261 1 was assumed. The measurements phased with the period 1.78 days show a single-wave variation while phasing of the magnetic field measurements with other periods mentioned in the literature did not display such a sinusoidal variability.

In this study of Hubrig et al. (2016), no longitudinal magnetic field measurement showed a detection at a significance level $\geqslant 3\sigma$. The highest significance levels in the measurements were achieved at rotational phase 0.037 (at 2.4σ) for the measurements using the entire spectral range and at rotational phase 0.668 (at 2.2σ) for the

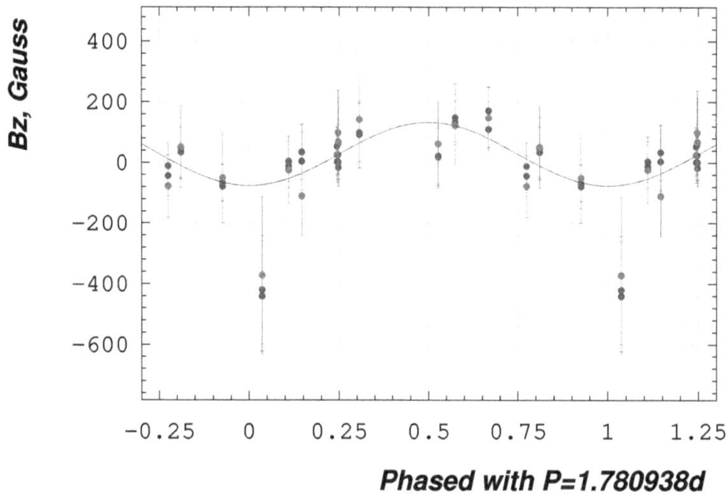

Figure 5.7. Longitudinal magnetic field variation of ζ Pup phased with the 1.78 days period reported by Howarth et al. (2014). Measurements of the magnetic field using absorption and emission lines are presented by red circles, those without emission lines by blue circles and those for hydrogen lines by green circles. The solid line represents a fit to the data. Credit: Figure 1 from Hubrig et al. (2016), © 2016. The American Astronomical Society. All rights reserved.

measurement carried out excluding the emission lines. The measurement obtained at phase 0.037 showed the highest field strength, but also the largest error. As the assigned weights are inversely proportional to the squares of the measurement errors in the fitting procedure, the contribution of this measurement is expected to be by a factor of ~7 lower than that for the other measurements. To check the likelihood of the single-wave variability of the longitudinal magnetic field, Hubrig et al. (2016) repeated the fitting procedure excluding this measurement. The result is presented in Figure 5.8, where the solid line shows the fit to the joined three sets of measurements. A rather similar mean value of the longitudinal magnetic field $\overline{\langle B_z \rangle} = 32.3 \pm 8.2$ G, a slightly smaller field amplitude $A_{\langle B_z \rangle} = 88.4 \pm 15.2$ G and a root mean square value of 46.7 G were obtained. Although no magnetic field measurements had a significance level higher than 2.4σ, the presence of a magnetic field cannot be completely excluded due to the single-wave variation of the longitudinal magnetic field measurements.

Intriguingly, the presence of a period of 1.78 days was also detectable in the variability of the equivalent widths of lines belonging to different elements. In Figure 5.9, the variations of the equivalent widths over the period of 1.78 days is presented. Clear variability was detected for all absorption lines studied. The equivalent widths of absorption hydrogen and helium lines decreased in the vicinity of the positive magnetic pole, while the intensity of the emission lines He II λ4686 and the blend of N II lines at ~4640 Å behaved opposite, showing an increase at the positive magnetic pole.

According to Howarth & van Leeuwen (2019), the runaway status, rapid rotation, and anomalous physical properties of ζ Pup are all indicative of an

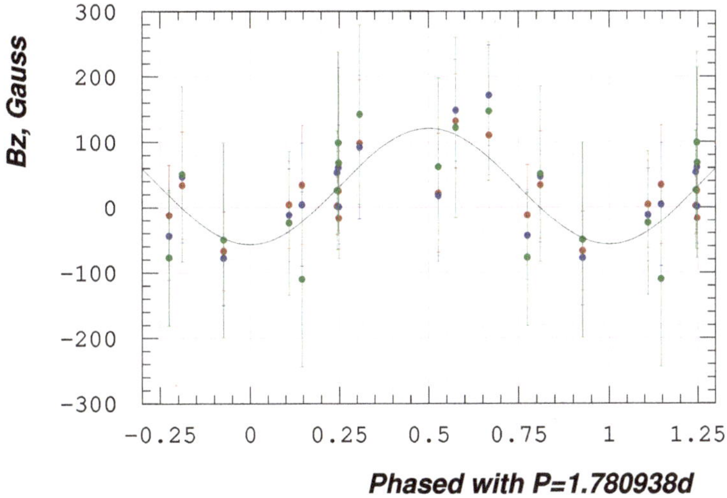

Figure 5.8. The same as in Figure 5.7, excluding the measurement with the lowest accuracy. Credit: Figure 2 from Hubrig et al. (2016), © 2016. The American Astronomical Society. All rights reserved.

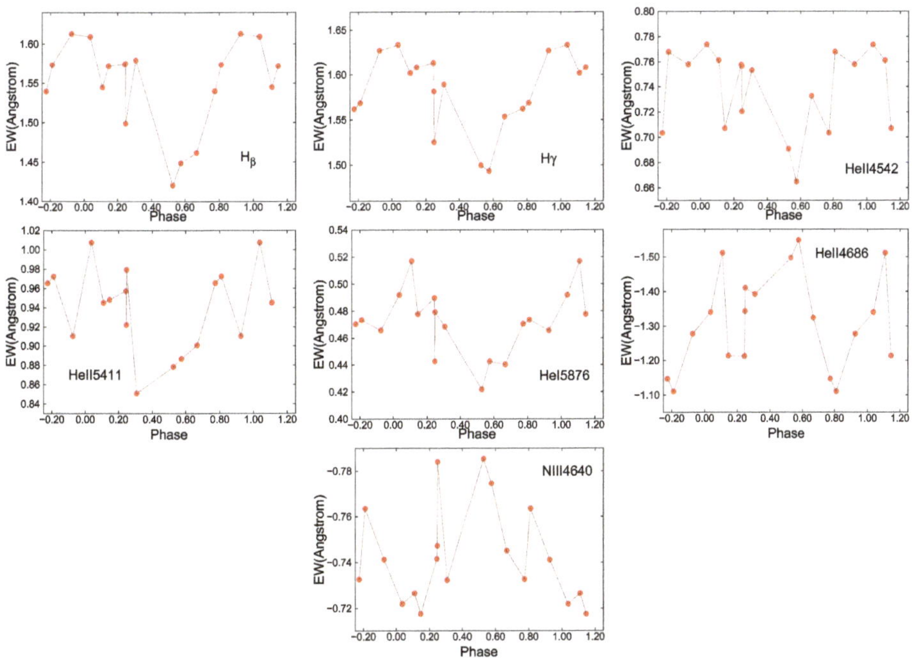

Figure 5.9. Variability of equivalent widths of lines belonging to different elements in the spectra of ζ Pup over the period of 1.78 days. The measurement accuracy corresponds to the size of the plotted symbols. Credit: Figure 6 from Hubrig et al. (2016), © 2016. The American Astronomical Society. All rights reserved.

evolutionary history involving binary (or multiple-star) interaction. Also the results of the kinematical studies by Hubrig et al. (2011c, 2011e, 2013) seem to indicate that the presence of a magnetic field is more frequently detected in candidate runaway stars. Another example of a massive star that is formed in a dramatic close binary interaction, i.e., in a merger of two stars or a dynamical mass transfer event is the fast rotating runaway O9.5V star ζ Oph with a bow-shock nebula. Figure 5.10 shows a multicolour image of ζ Oph based on archival Spitzer IRAC maps. Tetzlaff et al. (2010) reinvestigated the scenario of a binary supernova in Upper Scorpius involving ζ Oph and PSR B1929+10 and concluded that it is very likely that both objects were ejected during the same supernova event. ζ Oph has been extremely well studied in all wavelength ranges, from the X-ray to the infrared region with Spitzer. According to Kambe et al. (1993), it is possible that this star rotates with almost break-up velocity with $v \sin i = 400$ km s^{-1}. It is also known to exhibit non-radial pulsations

Figure 5.10. Combined IR Spitzer IRAC (3.6 μm blue, 4.5 μm green, 8.0 μm red) image of the bow shock around the runaway star ζ Oph from archival data. Galactic coordinates are shown. The image size is $\sim 36' \times 31'$. Credit: Figure 1 from Hubrig et al. (2011d).

with periods of hours, first discovered from ground-based observations (e.g., Kambe et al. 1997). Later, the star was monitored by the MOST satellite, which revealed at least a dozen significant pulsation frequencies between 1 and 10 day^{-1}, typical of β Cep-like pulsation modes (Walker et al. 2005). In IUE spectra, the UV resonance lines show multiple DACs (e.g., Howarth et al. 1984). Strong line-profile variability in optical spectra was reconciled with travelling sectorial modes of high degree (e.g., Reid et al. 1993).

The first detection of a mean longitudinal magnetic field in ζ Oph was announced by Hubrig et al. (2011d, 2011e, 2013) who used FORS 2 spectropolarimetric observations acquired over several years. The measured longitudinal magnetic field showed a change of polarity with the strongest field value of the order of −220 G at a significance level of 3.1σ. The rotation period of this star is still uncertain. The resulting periodogram for the magnetic field measurements using the entire spectra showed a dominating peak corresponding to a period of about 1.3 days. Figure 5.11 shows magnetic field variations determined from all lines phased with 1.3 days. Interestingly, Pollmann (2012) presented equivalent width changes of the He I λ6678 line with a period of 0.643 day, which is roughly half of the magnetic period of 1.3 days determined using the whole spectral region. For the rotation period of ~1.3 days the equivalent width of the He I line would then display a double wave, a behaviour of He I lines that is frequently found in He variable early-type Bp stars (Briquet et al. 2004). Clearly, additional future measurements are needed to confirm this periodicity of the magnetic field in ζ Oph.

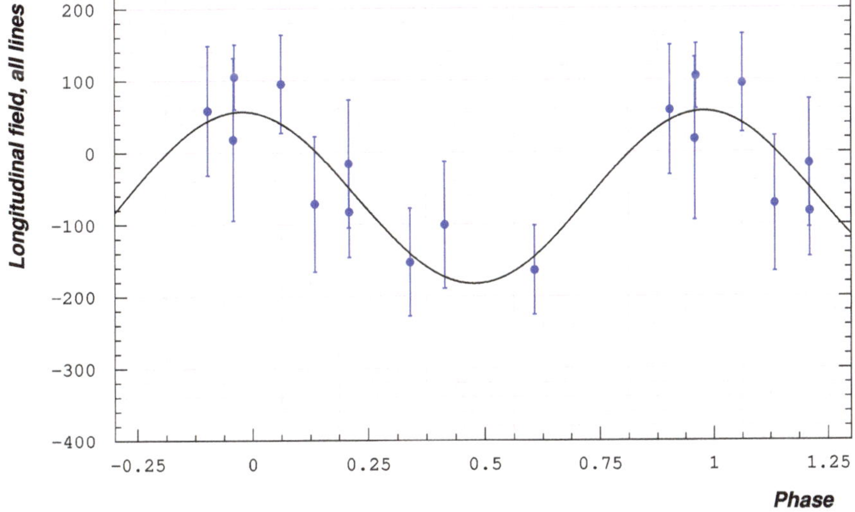

Figure 5.11. Phase diagram for the best sinusoidal fit corresponding to the period of 1.3 days for the longitudinal magnetic field measurements using the whole spectrum for ζ Oph. Credit: Hubrig et al. (2013), reproduced with permission © ESO.

The above discussed studies show that it is indispensable to test theoretical predictions to understand the role of magnetic fields in the physical processes taking place in stellar atmospheres of massive stars. Moreover, as currently only about a dozen O-type stars have confirmed magnetic fields, such test studies have also the potential to significantly increase the fraction of known magnetic O-type stars. Since the variability in massive stars on different timescales has diverse and sometimes uncertain origins, high signal-to-noise spectropolarimetric time series obtained with the shortest possible exposure times are extremely important to exclude or separate the impact of short-term variability on the magnetic field measurements as much as possible. For instance, as already discussed in Section 3.2, high-resolution spectropolarimetric observations of pulsating variable stars frequently fail to show credible measurement results, if the whole sequence of subexposures at different retarder waveplate angles has a duration comparable to the timescale of the pulsation variability.

5.2.2 Biases in the Studied Samples of O-type Stars

The reported detections of magnetic fields in massive stars with various spectral classifications indicate that magnetic fields are possibly present in stars with very diverse observed properties in UV, optical, and X-ray domains. As already mentioned above, all Galactic Of?p stars are magnetic and the presence of variable emission in the C III blend in these stars is indicative of circumstellar structure around them, related to their magnetospheres. However, the emission in the C III blend disappears entirely in late O-type stars (Walborn et al. 2010), meaning that such a discriminating emission effect cannot be used in cooler O-type stars. Obviously, the Of?p classification is very narrow, as it is limited to hotter stars with spectral types in the range O4f?p–O8f?p with effective temperatures between 34.5 and 41 kK. Due to the absence of specific classification criteria related to the presence of magnetic fields in cooler O stars, previous magnetic surveys predominately concentrated on the hotter and more massive O-type stars or on early B-type stars for which anomalously strong and variable He lines serve as direct classification criteria for their magnetic nature.

Among the 97 O-type targets observed in the framework of the MiMeS survey, only about one quarter have spectral types O8 to O9 with luminosity classes from V to III. In addition, the majority of O-type stars studied by the MiMeS consortium were rather young, located in the first half of the main sequence (Grunhut et al. 2017). Only eleven targets with spectral types O8 to O9 were observed within the framework of the BOB survey using HARPSpol high-resolution spectropolarimetry and 15 more with the low-resolution FORS 2 spectropolarimeter. Among the targets with O9 spectral classification, only two stars, HD 57682 and HD 54879, with spectral types O9.5IV and O9.7V, respectively, were identified as magnetic by the MiMeS and BOB consortia. And only two other magnetic targets with O8 spectral classification have been detected, the spectrum variable Of?p star HD 108 with O4–8f?p classification and HD 191612 with O6–8f?p classification. It appears that a significant increase of the sample of late O-type stars with measured circular polarisation is necessary to build trustworthy statistics on the occurrence of magnetic fields and the field strength distribution in O-type stars.

Furthermore, the location of magnetic massive O-type stars in the H-R diagram indicates that all of them have already evolved from the zero-age main sequence (ZAMS), implying that a certain amount of time is necessary for their magnetic fields to be built up and become measurable using spectropolarimetry (e.g., Castro et al. 2015). This suggests that observations of massive stars already evolved from the ZAMS bear a much larger potential to detect magnetic fields. The position of magnetic O and early-B stars in the H-R diagram is shown in Figure 5.12 (Castro et al. 2015). From this figure it is obvious, that in contrast to the location of the magnetic B-type stars, magnetic O-type stars appear significantly evolved from the ZAMS by approximately half of their main-sequence evolution time.

5.3 Magnetic Early-B Stars

The occurrence of magnetic fields slightly changes between the O and early-B spectral types with a higher occurrence of about 10% for early B-type stars. Currently, more than eight dozens of magnetic early B-type stars are known (e.g., Shultz et al. 2018). Similar to Ap/Bp stars, early-type Bp stars show photometric and spectroscopic variability caused by surface abundance spots of He, Si and various metals. The distribution of these spots is non-uniform and non-symmetric with

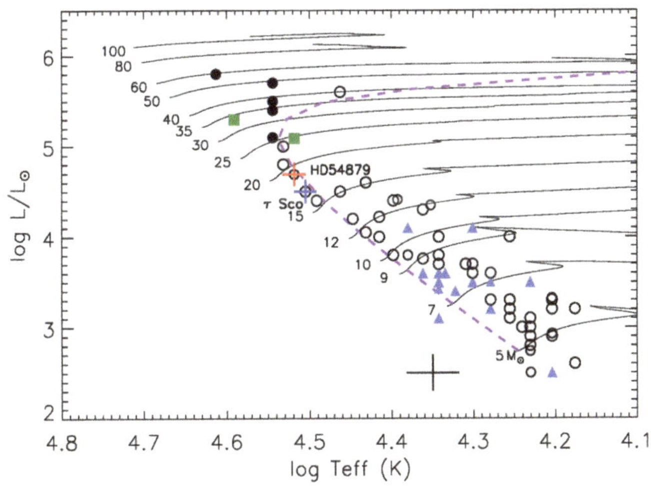

Figure 5.12. Position in the Hertzsprung–Russell diagram of OB stars with confirmed magnetic field detections and evolutionary tracks. Stars with an Of?p spectral classification are highlighted with filled black circles. Other peculiar O-type stars are indicated by green squares. The most recently discovered magnetic O-type star HD 54879 is labelled and shown with a red cross. The magnetic B0-type star τ Sco with an unusually complex, relatively weak surface magnetic field is also labelled and marked by a blue cross. Filled blue triangles indicate B-type stars classified as peculiar. An isochrone close to the age of HD 54879, t = 4.0 Myr, is plotted by a purple dashed line. The averaged uncertainties in effective temperature and luminosity are illustrated by the black cross in the bottom part of the plot. Credit: Castro et al. (2015), reproduced with permission © ESO.

respect to the rotation axis. Apart from the variability of the He and Si lines, also lines belonging to CNO and iron-peak elements frequently show variable line profiles over the rotation periods, but the distribution of these elements is poorly documented in the literature due to the relative weakness of their lines compared to the He and Si lines. Spots with He overabundance are usually concentrated toward one or both magnetic poles whereas metals are overabundant at intermediate latitudes, i.e., the lines of different elements with different abundance distributions across the stellar surface sample the magnetic field in different manners. Combining them together in the magnetic field measurements, as is frequently done with the LSD technique, may lead to a dilution of the magnetic signal or even to its cancellation, if enhancements of different elements occur in regions of opposite magnetic polarities.

An example of the He and Si spot distribution on the surface of a magnetic early-B star, the Doppler map calculated for the strongly magnetic B2 star CPD −57 3509 with a dipole strength in the range of 3.9–4.5 kG is presented in Figure 5.13 (Hubrig et al. 2017c). For this star, application of the Doppler imaging technique to high-resolution spectra obtained with the UV-Visual Echelle Spectrograph (UVES; Dekker et al. 2000) shows the presence of He spots located around the magnetic poles, with a strong concentration at the positive pole and a weaker one around the negative pole (Hubrig et al. 2017c). In contrast, high abundance Si spots are located close to the magnetic equator.

Among the previously studied magnetic early-B stars, the record holder is the He-rich star CPD −62 2124 with a mean longitudinal magnetic field $\langle B_z \rangle$ up to 6.8 kG detected using the hydrogen lines and $\langle B_z \rangle = 6.3$ kG using the entire spectrum with all available spectral lines in the polarimetric spectra obtained with FORS 2 (Castro et al. 2017; Hubrig et al. 2017b). Among the known magnetic massive stars, only NGC 1624-2 with a $\langle B_z \rangle$ of about 4.5 kG (e.g., Järvinen et al. 2021) and the early-B star HD 64740 with $\langle B_z \rangle \approx 4.8$ kG (Petit et al. 2013) host a magnetic field with a strength comparable to that of CPD −62 2124. The mean longitudinal magnetic field measurements using the hydrogen lines or those using the entire spectrum phased with the rotation period of 2.628 days determined from photometric, spectroscopic and spectropolarimetric observations are shown in Figure 5.14. The presented $\langle B_z \rangle$ phase curve displays a secondary minimum at phase 0.5, indicating the possibility of a more complex magnetic field with a dominant dipole component and a non-zero quadrupole component (Hubrig et al. 2017b). Using the relations for a centred magnetic dipole tilted to the rotation axis (see Chapter 2), the estimated dipole strength B_d for this star is about 21 kG.

Similar to strongly magnetic, slowly rotating Ap/Bp stars, a number of early-B stars with strong magnetic fields show resolved Zeeman split spectral lines. Studies of such stars permit to set additional constraints on the magnetic field geometry by measuring the mean magnetic field modulus. Examples of magnetically split lines in the three early-B stars HD 58260, HD 96446, and HD 149277 (Hubrig et al. 2019a) are presented in Figure 5.15. The B2IV/V star HD 149277 with a mean magnetic field modulus of 9.0 kG is a rare close SB2 system with a slowly rotating magnetic He-rich primary with $P_{rot} = 25.4$ days determined using mean longitudinal magnetic

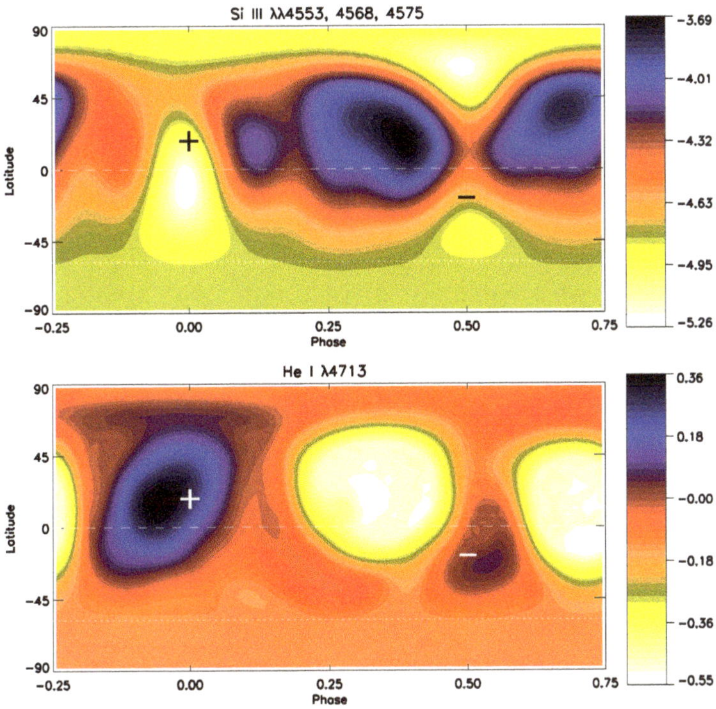

Figure 5.13. Mercator abundance maps plotted together with the location of the magnetic poles for CPD −57 3509. The distribution of Si is shown at the top and that of He at the bottom. The plus-sign denotes the location of the positive magnetic pole, and the negative magnetic pole is indicated by the minus-sign. The dashed line shows the stellar rotational equator, and the dotted line represents the limit below which the stellar surface is not visible due to the rotation axis inclination with respect to the line of sight. Note that the rotational phases range from −0.25 to 0.75 to better display the area around the magnetic poles. Credit: Figure 7 from Hubrig et al. (2017c).

field measurements (Shultz et al. 2018). Using high-resolution polarimetric spectra of HD 149277 acquired with ESPaDOnS at the Canada–France–Hawaii Telescope, Shultz et al. (2018) reported on the presence of a rather strong mean longitudinal magnetic field with $\langle B_z \rangle_{max} = 3.3 \pm 0.1$ kG. To estimate the orbital parameters of the SB2 system, González et al. (2018) used ESPaDOnS spectra recorded from 2012 to 2016. The analysis of these spectra revealed $P_{orb} = 11.519\,2 \pm 0.000\,5$ day indicating extremely strong subsynchronous rotation of the primary component. Such a strong subsynchronous rotation was not detected in any other binary system with a magnetic chemically peculiar component and is most likely caused by magnetic braking. The rotational variability of the resolved Zeeman split Fe III λ5073.6 line in the primary component is presented in Figure 5.16. As illustrated in this figure, the Zeeman split Fe III line is in certain rotation phases asymmetric about the line centre with the blue component appearing less deep than the red component, and with the blue component deeper in other phases. Similar asymmetries occur in the same way in other transitions, indicating that the rotational Doppler effect is non-negligible, i.e., various parts of the stellar surface characterised by different magnetic field

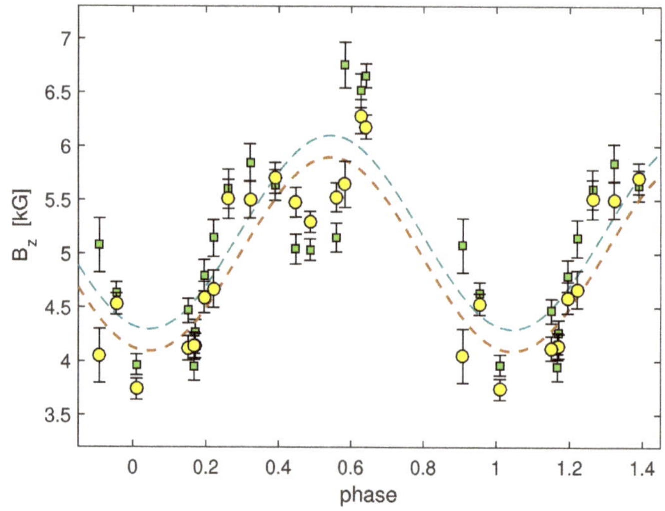

Figure 5.14. Phase curves for the $\langle B_z \rangle_{\mathrm{all}}$ and $\langle B_z \rangle_{\mathrm{hyd}}$ magnetic measurements (yellow circles and green squares, respectively) for the strongly magnetic early B-type star CPD −62 2124. The sinusoidal fits obtained for the measurements using the entire spectra and those using hydrogen lines are presented by the lower and upper dashed lines, respectively. Credit: Figure 5 from Hubrig et al. (2017b).

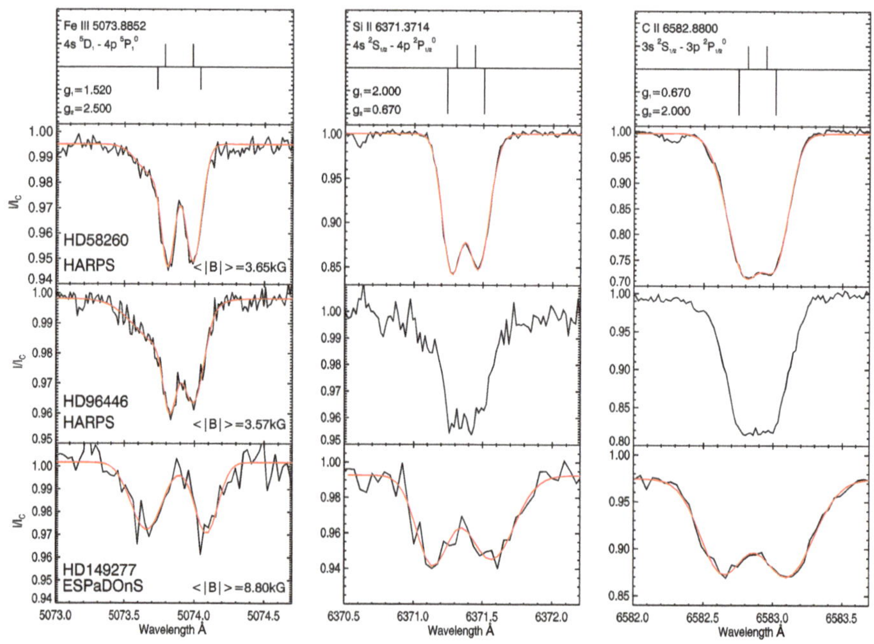

Figure 5.15. The magnetically split lines Fe III λ5074, Si II λ6371, and C II λ6583 in high-resolution Stokes *I* spectra of HD 58260 with a mean magnetic field modulus of 3.65 kG, HD 96446 with a mean magnetic field modulus of 3.57 kG, and HD 149277 with a mean magnetic field modulus of 8.80 kG. The red lines denote the fit of a double-Gaussian to the data. For two lines in the spectrum of HD 96446, Si II λ6371, and C II λ6583, the splitting is not sufficient to allow a proper fit. Credit: Figure 1 from Hubrig et al. (2019a).

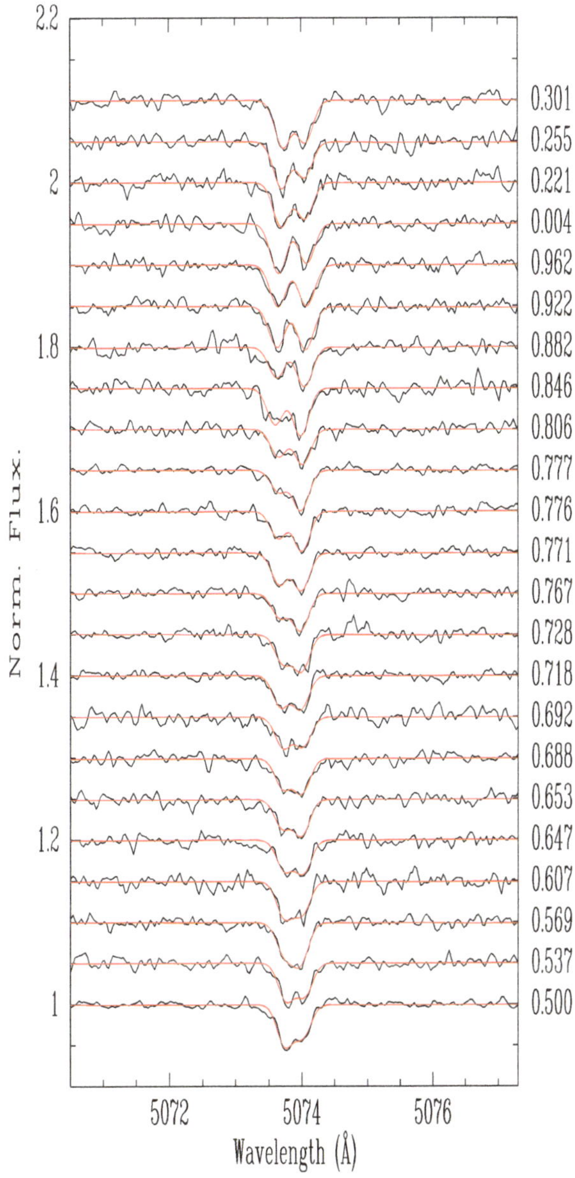

Figure 5.16. Variability of the magnetically split Fe III λ5073.6 line over the rotation period in the disentangled spectra of the primary component in the system HD 149277. Spectra are vertically offset for better visibility and are sorted in rotational phase (see right side; starting at 0.5) from bottom to top. The red lines represent a double-Gaussian fit to the data. Figure 2 from González et al. (2018).

strengths contribute to the line profile. Further it was found that in the primary star, the maximum of the magnetic field modulus coincides roughly with the positive extremum of the longitudinal magnetic field, whereas the minimum of the field modulus is coinciding with the negative extremum of the longitudinal magnetic field.

No evidence for a longitudinal magnetic field was seen in the circularly polarised spectra of the secondary component (González et al. 2018).

In a number of magnetic Bp stars, apart from the rotational variability, short-term photometric and spectroscopic variability is observed due to the presence of coherent p and/or g pulsation modes (e.g., González et al. 2019) triggered by an opacity mechanism operating in the Z-bump relating to iron-peak elements (e.g., Dziembowski et al. 1993; Miglio et al. 2007). For the well-known He-rich B2p star HD 96446 with a strong, kG order longitudinal magnetic field, the discovery of radial velocity variations with a period around 2.23 h attributed to β Cep p-mode pulsations was reported by Neiner et al. (2012). More recently, its pulsational and rotational variability was studied in great details using extensive spectroscopic time series by González et al. (2019). The authors employed an immense data set consisting of 646 spectra gathered with the spectrograph CORALIE (Queloz et al. 2000) mounted on the Swiss 1.2 m Leonhard Euler Telescope in La Silla (Chile). This data set was completed by spectra obtained with five other spectrographs to identify the rotation period of 23.4 days and two pulsational modes with periods of 0.092 92 and 0.110 8 day. Further, the authors discovered that HD 96446 is a SB2 system with an eccentric orbit ($e = 0.59$) with $P_{\rm orb} = 799$ days. Its primary shows an inhomogeneous distribution of chemical elements over the stellar surface. The mass ratio was found to be very low with $q = 0.29$, resulting in a visual light-ratio of the components of only ~0.05. Spectropolarimetric observations indicated that helium is concentrated at the negative magnetic pole and most metals accumulate at lower latitudes.

Especially if the length of the spectropolarimetric subexposures with different orientation of the quarter-wave retarder plate amounts for a fraction of the pulsation period, the presence of radial velocity shifts due to the pulsational variability generates spurious signatures and a distortion of the Stokes V and the diagnostic null spectra that are used in the LSD technique to check for spurious effects in the field measurements. The pulsational line-profile variability, in addition to radial velocity shifts, also has a significant impact on the line-profile shape (see Figure 3 of Hubrig et al. 2011a, and Section 3.2). Hubrig et al. (2011a) and Järvinen et al. (2017) presented in their work a specific measurement strategy to account for pulsational variability.

Shultz et al. (2018) presented a compilation of published and newly discovered magnetic stars with spectral types between B5 and B0 with sufficient spectropolarimetric data to study their magnetic characteristics. In contrast to Ap/Bp stars, massive Bp stars showed a higher percentage of rapidly rotating stars. Within the studied sample of about 50 stars, the median rotation period was of the order of one day. The mean longitudinal field phase curves obtained for the majority of the studied stars were consistent with dipolar magnetic fields and only twelve stars showed a contribution of higher order multipoles. A comparable fraction of magnetic stars possessing magnetic phase curves fitted by a double wave was reported by Hubrig et al. (2007) for a sample of late type Bp stars. The study of the evolution of the magnetic field strength and magnetic flux on the main sequence based on the same sample of massive Bp stars (Shultz et al. 2019) showed very similar results compared to the results obtained by Hubrig et al. (2007) for magnetic late-B stars with masses $\leqslant 6\ M_\odot$, who found that the distribution of the relative ages peaks at the ZAMS and the strongest magnetic fields are usually detected in faster rotating and younger stars in terms of the elapsed fraction of their main-sequence life. The fact that the strongest magnetic fields are observed only close to the ZAMS can be interpreted as a magnetic field decay in stars at advanced ages.

According to Shultz et al. (2019), rotation in massive early B-type stars can slow down over time, which is qualitatively consistent with magnetic braking. Magnetic braking for B-type stars can be strong enough to become directly observable by analysing spectroscopic, photometric, and magnetic field observations. A number of stars were suggested to show variations of their rotation periods. Comparing long baseline photometric data, an increasing rotation period has been detected for the magnetic B-type star σ Ori E (Townsend et al. 2010). For a number of stars, even decreasing rotation periods were reported. As an example, Mikulášek et al. (2007) discovered possible gradual changes in the stellar rotation of HD 37776, a very young B2IV star. However, later on, it was found that the rotational deceleration was replaced by acceleration (Mikulášek et al. 2008, 2011). Currently, there is no explanation for this type of rotational variability.

5.4 Magnetospheres

Magnetic fields of massive stars can impact the surface rotation rates via magnetic braking (Weber & Davis 1967; ud-Doula et al. 2008), introduce chemical abundance inhomogeneities and peculiarities (Hunger & Groote 1999), and confine the stellar wind in a magnetosphere (e.g., Friend & MacGregor 1984; Babel & Montmerle 1997b; ud-Doula & Owocki 2002). Therefore, these stars are also unique targets for the study of stellar magnetospheres. The magnetic confinement forces most of the wind to be driven along the field lines toward the magnetic equator. In the vicinity of the equatorial plane the wind components of the two hemispheres collide leading to strong shock and thereby heating plasma to X-ray temperatures (Babel & Montmerle 1997b). The magnetically confined wind shock (MCWS; Babel & Montmerle 1997a, 1997b) model describing the trapping and channelling of the stellar wind in closed magnetic loops is, in combination with rotation effects, usually used to explain the observed X-ray emission in magnetic massive stars (e.g., ud-Doula & Nazé 2016).

The impact of a dipole magnetic field rooted in the star on the stellar wind is characterised by the confinement parameter (ud-Doula & Owocki 2002)

$$\eta_* = \frac{B_{\rm eq}^2 R_*^2}{\dot{M} v_\infty}, \tag{5.1}$$

where $B_{\rm eq}$ is the magnetic field strength at the equator, R_* the stellar radius, $\dot{M}$ the mass loss rate, and v_∞ the terminal wind speed.

The most obvious signature of the presence of a circumstellar magnetosphere around massive stars is the presence of variable emission in hydrogen lines. The magnetospheres of massive stars are classified as dynamical or centrifugal magnetospheres depending on whether rotational support plays a negligible or dominant role in shaping the corotating, magnetically confined stellar wind (Petit et al. 2013). Corotation can lead to significant centrifugal forces acting within the inner magnetosphere. At the Kepler corotation radius R_K, gravitational and centrifugal forces are equivalent (e.g., Townsend & Owocki 2005; ud-Doula et al. 2008). On the other hand, the Alfvén radius R_A marks the point where the gas flow becomes dominant over the magnetic field (e.g., Küker 2017). For slowly rotating stars with $R_A < R_K$, material trapped in closed magnetic loops falls back to the star on a dynamical timescale, forming a dynamical magnetosphere (DM; e.g., Sundqvist et al. 2012). For the more rapidly rotating case with $R_A > R_K$, material caught in the region between R_A and R_K is centrifugally supported against infall, and so builds up to a much denser centrifugal magnetosphere (CM). Slowly rotating magnetic O stars with a narrow Hα emission are consistent with a DM and more rapidly rotating magnetic early-B stars with a broader Hα emission are associated with a CM.

Figure 5.17 illustrates simulations using the Nirvana MHD code (Ziegler 2008, 2011) using stellar parameters of the magnetic O9.7V star HD 54879 (Shenar et al. 2017; Hubrig et al. 2019b) with a setup similar to that described by Küker 2017.

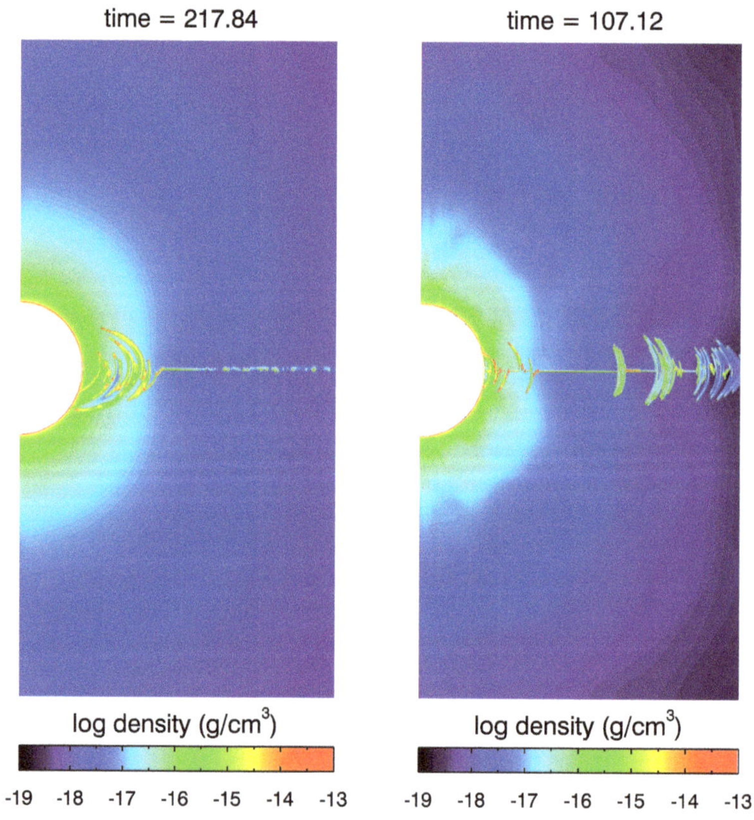

Figure 5.17. Snapshots from simulations of a line-driven wind and a dipole magnetic field rooted in the magnetic O9.7V star HD 54879. The polar field strength is 100 G (*left panel*) and 300 G (*right panel*). Colours indicate mass density. The circular cut-out at the left margin of the plots marks the star. The left margin is the dipole axis of the magnetic field. Credit: Figure 10 from Hubrig et al. (2019b).

As simulations for a 2 kG order polar field strength proved to be very time consuming, only the cases for 100 and 300 Gauss were considered. The simulation box is a spherical shell with the stellar surface as the inner boundary. The outer boundary is typically at a radius of ten stellar radii. The magnetic field is kept fixed on the inner boundary and can evolve elsewhere. The geometry is axisymmetric with respect to the polar axis of the magnetic field and the gas is assumed isothermal, with the gas temperature being the effective temperature of the star. At the outer boundary, the Nirvana code's built-in outflow conditions are used. Likewise built-in boundary conditions are used on the symmetry axis.

It is obvious that the magnetic field has a profound impact on the gas flow. The 100 G case shows a fully-developed magnetosphere while in the 300 G case the magnetosphere is still forming. Stellar rotation was not included, hence the dipole axis of the magnetic field defines the symmetry. Outside the confinement radius R_K, gas still concentrates in elongated structures along the magnetic field lines, but is ejected away from the star in episodic outbursts. While for very weak magnetic fields having a small value of the confinement parameter the mass loss is steady, these simulations show episodic outbreaks of gas previously trapped near the magnetic equator. These mass ejections are typical for the case of strong confinement, which is fulfilled even for the weakest field of 100 G. During the outbursts, which last for about a day, clumps of gas move away from the star in the equatorial plane of the magnetic field at high speed. The contribution of these outbursts to the total mass loss rate reaches values of the same order as that of the steady wind component, i.e., the total mass loss rate during the outburst can be twice as high as during the quiet phase in between. This is in contrast to cases of weak confinement, $\eta_* < 1$, where a steady disk forms and the mass loss rate shows little variation with time.

The very interesting case of the star NGC 1624-2 with the strongest magnetic field among the known magnetic O-type stars and a giant magnetosphere was discussed in Section 5.2. NGC 1624-2's magnetosphere was estimated to extend to an Alfvén radius R_A of more than eleven stellar radii, hence trapping 95% of the outflowing wind, much more than other magnetic O stars that have an R_A of just a few stellar radii (David-Uraz et al. 2019). For this star with $B_d \geqslant 16$ kG, the variable Stokes V signatures observed in the emission He I lines correspond to a longitudinal magnetic field of negative polarity while the inverted Stokes V signatures observed in high-excitation emission O III lines show a positive longitudinal magnetic field. Such observations suggest a very complex structure of the magnetospheric circumstellar environment of NGC 1624-2.

Using the Nirvana MHD code in a similar setup as for HD 54789, Järvinen et al. (2021) presented the first MHD numerical simulations of the mass distribution and gas motion in the magnetosphere of this strongly magnetic star. To limit the cost in CPU time, the outer boundary of the simulation box was placed at 5 stellar radii and a polar field strength of 4 kG was chosen. While this was smaller than the observed field strength of 16 kG, it still leads to an extended region of magnetic confinement with $\eta_* \approx 700$ and an Alfvén radius of 5.4 stellar radii. Within the simulation box, the magnetic field geometry stayed close to the original dipolar geometry and only a fraction of the gas flowing from the stellar surface could escape the system. In the initial phase of the simulation run, the gas flow and mass density underwent drastic changes. While the gas density at high latitudes dropped, gas piled up in the equatorial plane, where a thin disk formed from which filaments grew that were aligned with the magnetic field. Figure 5.18 shows a snapshot of the mass density taken at a time when the system has reached a semi-stationary state insofar as the general mass distribution with a high mass concentration at low to mid-latitudes did not evolve much anymore. The right panel of Figure 5.18 shows a zoom-in to the inner part of the simulation box. The shape and length of the individual filaments is highly variable, though, and the system does not reach a truly stationary state. All the shown field lines are closed. Compared to the original dipole, the field became slightly compressed by the gas load but the general field topology remained unchanged. This is generally different for weak surface fields or at greater distances from the star, where the field lines are stretched away from the star and a thin disk of outflowing material forms. For the case of a rather strong magnetic field presented in

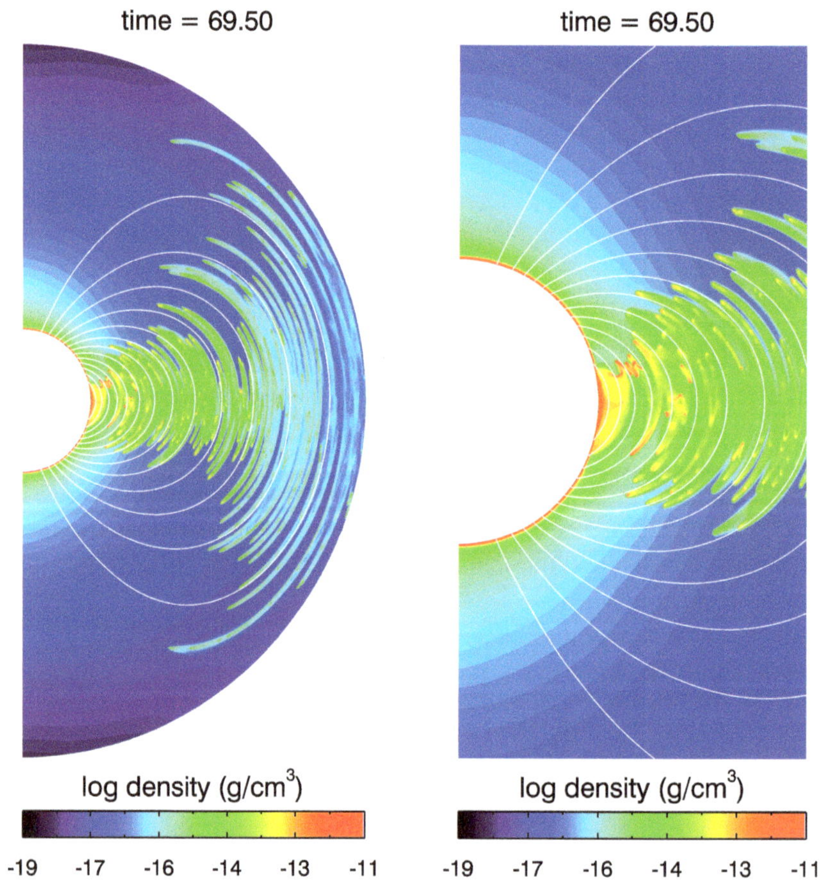

Figure 5.18. Mass density and magnetic field structure in the vicinity of the strongly magnetic O7?p star NGC 1624-2. *Left panel*: the whole simulation box. *Right panel*: zoom into the central region. The unit of the time stamps is ks. Credit: Figure 7 from Järvinen et al. (2021).

Figure 5.18, the regions of the stellar surface where open field lines originate is limited to a small range of latitudes around the poles. It is expected that this effect should even be stronger with a dipole field strength as large as 16 kG.

5.5 RRM Stars

The strong magnetic field and fast rotation in some early B-type stars lead to the formation of a centrifugally supported magnetosphere with rotationally modulated hydrogen-line emission, explained within the rigidly rotating magnetosphere (RRM) model (e.g., Townsend et al. 2005). This model predicts the accumulation of circumstellar plasma in two corotating clouds, situated in magnetohydrostatic equilibrium at the intersections between the magnetic and rotational equators. It is essentially a generalised extension of the Babel & Montmerle (1997b) model, presenting a semianalytic method for modelling the circumstellar environment of early-type stars and assuming that the magnetic field lines remain rigid and corotate with the star without being influenced by the dynamics of the wind.

Early B-type stars with strong magnetic fields and very fast rotation are extremely rare. The emission variability in hydrogen lines observed in the spectra of σ Ori E suggests that gas is trapped in magnetospheric clouds (e.g., Groote & Hunger 1997). Currently, only four other stars, HR 5907 (=HD 142184), HR 7355 (=HD 182180), HD 23478, and HD 345439 are known to belong to this group, which is called the σ Ori E, analogues (e.g., Grunhut et al. 2012; Rivinius et al. 2008).

The latest discovery of the presence of an RRM in early B-type stars was reported using high-resolution ($R \approx 22,500$) near-infrared H-band spectra of HD 23478 and HD 345439 obtained in the Apache Point Observatory Galactic Evolution Experiment (APOGEE; Eikenberry et al. 2014). These stars displayed prominent Brackett-series emission lines with a characteristic double-horned profile. The type of profile and the peak separation appeared to be typical for the RRM features previously discovered in the early-B star σ Ori E, which possesses a strong magnetic field and rotates fast. Near-IR spectra of these two σ Ori E analogues, together with a spectrum of σ Ori E, obtained with the TripleSpec instrument (Wilson et al. 2004) are presented in Figure 5.19.

Wisniewski et al. (2015) analysed multi-epoch photometric observations of HD 345439 from the Kilodegree Extremely Little Telescope, Wide Angle Search for Planets, and ASAS surveys and detected in each data set the presence of a ~0.770 18 day rotation period. The authors suggested that the He-strong star HD 345439 of spectral type B2V is one of the fastest known He-strong σ Ori E analogues along with HR 7355 ($P_{rot} = 0.52$ day; Rivinius et al. 2008) and HR 5907 ($P_{rot} = 0.51$ day; Grunhut et al. 2012). Follow-up observations of HD 23478 and HD 345439 using low-resolution FORS 2 polarimetric observations revealed the presence of a strong longitudinal magnetic field of kG order in both stars (Hubrig et al. 2015a, 2017a). FORS 2 magnetic field measurements carried out for HD 345439, those using the entire spectra and those using only the hydrogen lines, phased with the rotation period of 0.770 18 day are presented in Figure 5.20. The obtained measurements suggested a lower limit for the dipole strength of $B_d \geqslant 8.98 \pm 0.49$ kG.

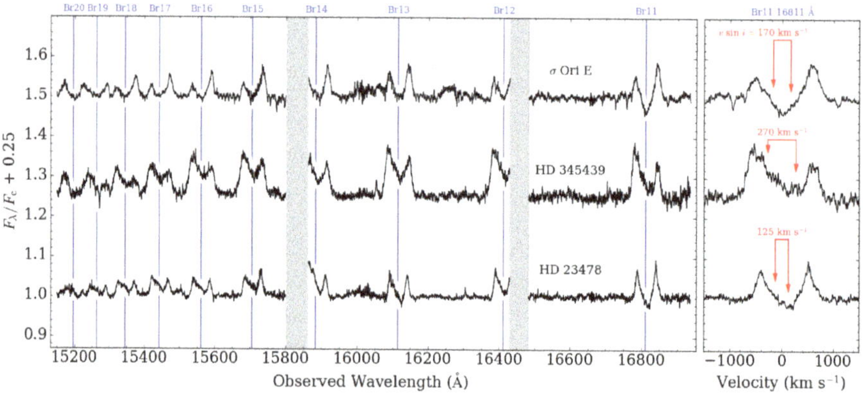

Figure 5.19. Portions of the normalised TripleSpec NIR spectra of HD 345439 (*middle*), HD 23478 (*bottom*), and the RRM archetype σ Ori E (*top*). The strongest features of all three spectra are the extremely broad double-horned Brackett-series emission profiles that distinguish the rigidly rotating magnetospheres of these stars. Credit: Figure courtesy of Drew S. Chojnowski.

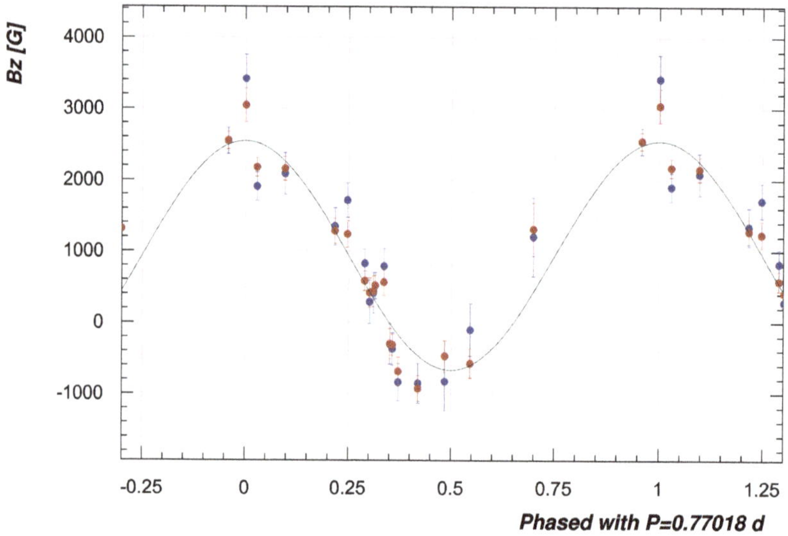

Figure 5.20. Longitudinal magnetic field variation of HD 345439 phased with the 0.770 18 day period. Measurements of the magnetic field using the entire spectrum are presented by red circles and those using hydrogen lines by blue circles. The solid line represents a fit to the data with a mean value for the magnetic field of $\overline{\langle B_z \rangle} = 939 \pm 96$ G and an amplitude of $A_{\langle B_z \rangle} = 1607 \pm 99$ G. For the presented fit, a zero phase corresponding to a positive field extremum at MJD56 925.542 5 was assumed. Credit: Figure 2 from Hubrig et al. (2017a).

In Figure 5.20, a noticeable dispersion of the measurements indicating a deviation from the simple dipole model is observed around the rotational phase 0.4. This may point to a more complex topology of the magnetic field structure in this star. On the other hand, the largest dispersion in the hydrogen equivalent width measurements appeared around the same phase range and can likely be explained assuming an occultation by circumstellar gas clouds magnetically confined to the magnetic equatorial plane (e.g., Hubrig et al. 2015b).

The rotation period of HD 23478 of about 1.05 days was determined by Sikora et al. (2015) using the Hipparcos Epoch Photometry catalogue (Perryman et al. 1997). Comparisons with the rigidly rotating magnetosphere model suggested that the photometric variability of HD 23478 may be caused by variable occultation of the stellar disc by the magnetosphere (Townsend & Owocki 2005). A surface dipole magnetic field of strength $B_d \geqslant 9.5$ kG was estimated using high-resolution ESPaDOnS, Narval, and median-resolution DimaPol spectropolarimetric observations (Sikora et al. 2015).

Very strong magnetic fields and extraordinary fast rotation of stars with rigidly rotating magnetospheres provide a mystery for theories of both star formation and magnetic field evolution. All five stars have similar dipole strengths of the order of 10 kG and short rotation periods. However, the spin-down timescale via magnetic braking is expected to be much shorter than their estimated ages (e.g., Rivinius et al. 2013; Mikulášek et al. 2010). The understanding of the RRM stars is further hampered by the substantial uncertainty regarding their rotational and magnetic properties. The studies by Hubrig et al. (2007) and Shultz et al. (2018) indicated that the magnetic topology in Bp stars is more complex than in lower mass Ap stars. Also Das et al. (2019) concluded that these extremely fast rotating magnetic stars do not show ideal dipolar magnetic fields, so that it is very important to have precise constraints on their surface magnetic structure. The topology of the magnetic field was reported to play a crucial role in the case of σ Ori E, for which the presence of a strong quadrupolar component was suggested by Leto et al. (2012). The authors searched for coherent emission at frequencies below 15 GHz using new and archive radio measurements obtained with the Very Large Array and concluded that the presence of a quadrupolar component of the magnetic field, dominant within a few stellar radii from the star, where the maser emission should be generated, inhibits the onset of the cyclotron maser instability.

The presence of remarkably strong magnetic fields controlling the stellar winds and fast rotation makes these type of stars a test case for models of X-ray production. Only low-resolution X-ray spectra were so far recorded for three members of this group, σ Ori E, HR 5907, and HR 7355 (e.g., Petit et al. 2013; Nazé et al. 2014). The analysis of archival XMM-Newton and Chandra observations of HR 5907 and HR 7355 showed that these stars are even harder X-ray sources than σ Ori E, with an emission measure weighted temperature of the hot plasma of 80 MK in HR 5907 and 60 MK in HR 7355 (Hubrig et al. 2015a). It is of interest that such high temperatures are very difficult to explain in the framework of the magnetically confined wind shock model suggested by Babel & Montmerle (1997b).

5.6 Space Photometry: Long- and Short-term Variability

Due to the presence of magnetospheres around magnetic O and early-B stars, periodic variability is frequently observed at different wavelengths, from the ultraviolet and X-ray spectral regions to the near-infrared. The availability of high-precision photometric light curves of massive stars obtained with the space missions CoRoT (Convection, Rotation et Transits planétaire; Auvergne et al. 2009), Kepler K2 (Borucki et al. 2010; Koch et al. 2010; Howell et al. 2014) and *TESS* (Transiting Exoplanet Survey Satellite; Ricker et al. 2015) is of immense importance to investigate the effect of magnetism and has allowed astronomers to obtain a wealth of information on the variability of massive stars. Periodic photometric variability associated with rotational modulation was already reported for a number of already known magnetic stars (Barron et al. 2020; Burssens et al. 2020). Moreover, depending on the type of variability, it is possible to identify magnetic candidates directly from their light curves (e.g., Pedersen et al. 2019; Burssens et al. 2020). The detection of rotational modulation of the light curves is necessary to support physical modelling of the spectroscopic and magnetic field variations.

High-precision short-cadence photometric observations with satellites also permit to explore the domain of magneto-asteroseismology. Pulsations in massive stars are commonly associated with coherent pulsation modes triggered by an opacity mechanism operating in the Z-bump related to iron-peak elements (e.g., Dziembowski et al. 1993; Miglio et al. 2007; see Section 7.5 for more details). According to Blomme et al. (2011), early O-type stars typically show stochastic low-frequency variability (SLF), whereas late O- and early B-type stars usually show coherent pulsation modes. It was suggested that the transition between these types of variability is at a spectral type around O8. Burssens et al. (2020) studied the variability of 98 massive O and B stars using TESS Sectors 1–13 observations and reported the possible detection of coherent p- and g-modes in a number of main sequence stars with spectral types in the range O4.5–O9.5. Bowman et al. (2020) explained the ubiquitous detection of stochastic variability in the photospheres of massive stars by the presence of Internal Gravity Waves (IGW). Due to the large uncertainties in the knowledge of the different physical processes taking place in the atmospheres and interiors of these stars, a number of ingredients in current stellar atmosphere and stellar evolution codes remain missing.

High-cadence TESS data are also promising for studies of the pulsational variability of magnetic O-type stars, which usually have long rotation periods of the order of months or years. The sample studied by Burssens et al. (2020) contained three main-sequence O-type stars with confirmed magnetic field detections, HD 37022, HD 54879, and HD 57682. However, coherent pulsation modes were not detected in these stars. TESS 2-min cadence observations of this remarkable magnetic star, also known as TIC 354879267, taken in Sector 19, consist of data spanning 25.06 days and comprising 16 920 data points. TESS observations of the slowly rotating O7f?p star NGC 1624-2 are of special interest as this star possesses the strongest magnetic field detected among O-type stars and a giant magnetosphere. The light curve is presented in Figure 5.21 in the bottom panel, revealing a peak-to-peak scatter of less than 0.04 mag. The variability that is evident to the eye is either instrumental, or non-periodic, or both. As can be seen in Figure 5.21 in the upper panel, within the 0.04-mag spread of the data an amplitude spectrum reveals three

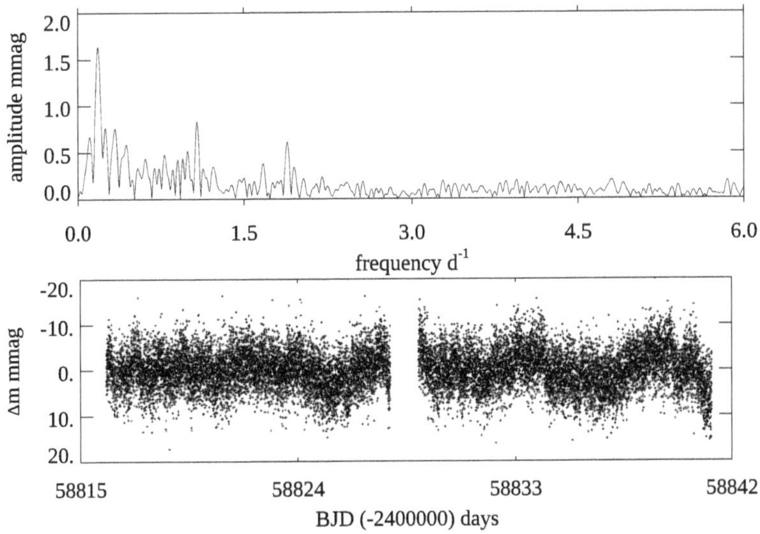

Figure 5.21. TESS light curve (*lower panel*) and the periodogram (*upper panel*) of the strongly magnetic O7f?p star NGC 1624-2. Credit: Figure 1 from Kurtz et al. (2020), © 2020. The American Astronomical Society. All rights reserved.

significant, coherent low-frequency peaks at $\nu = 0.187\,8$ day^{-1} ($A = 1.69$ mmag, S/N = 36), $\nu = 1.081\,0$ day^{-1} ($A = 0.79$ mmag, S/N = 17), and $\nu = 1.886\,8$ day^{-1} ($A = 0.63$ mmag, S/N = 14). The detected peaks most probably correspond to g-modes, as they are in the same frequency range as the g-modes identified in six early B stars by Burssens et al. (2020). No evidence of any pulsation was found at higher frequencies. It is of interest that a study of the magnetic field geometry of NGC 1624-2 by Järvinen et al. (2021) indicated the presence of line-profile variability on a timescale of several days, possibly related to the g-mode pulsations detected in TESS observations (Kurtz et al. 2020).

Currently, NGC 1624-2 is the only magnetic O star for which coherent pulsation modes have been detected. Importantly, this star is also intrinsically luminous and variable in X-rays. Petit et al. (2015) reported that the emergent X-ray flux is about 30% lower at the best visibility of the magnetic equator compared to the measured X-ray flux in the vicinity of the magnetic pole, indicating the presence of a rotational modulation of the X-ray flux. It is not clear whether NGC 1624-2 also shows pulsational variability in the X-ray flux, similar to that detected in the magnetic early-B star ξ^1 CMa, which is a typical β Cep pulsator with a period of 0.209 6 day. The broadband X-ray flux of ξ^1 CMa was found to exhibit a 10% modulation on the same period and in phase with the non-radial pulsations (Oskinova et al. 2014). Since the mechanism through which the X-rays are affected by stellar pulsations remains enigmatic, a search for X-ray pulsations in the strongly magnetic NGC 1624-2 will help to improve our understanding of processes taking place in the stellar magnetospheres and the connection between the pulsations at the level of the photosphere and the X-ray emission generated in the stellar wind.

5.7 Magnetic Field Origin and Evolution of the Field Strength

The origin of magnetic fields in massive stars is still unknown. Theoretic studies offer three explanations for their origins: fossil fields inherited from the star-forming cloud (e.g., Moss 2001), dynamo action taking place in the rotating cores of main sequence stars (Charbonneau & MacGregor 2001) and field generation in mergers of pre-main sequence or main sequence stars (Ferrario et al. 2009; Wickramasinghe et al. 2014). The first two hypotheses are unable to explain why only a small subset of massive stars is magnetic in spite of the fact that some of them form from the same molecular cloud and are governed by the same physical processes. They also do not explain the low (~10%) occurrence of magnetic stars and their broad range of field strengths.

To answer the principal question of the possible origin of magnetic fields in massive stars, it is important to build trustworthy statistics on the occurrence of magnetic fields and their topology at different evolutionary stages. However, no volume-limited study was presented yet. The observed stellar samples greatly suffered from a number of biases in the sample selection related to specific spectral classifications. The surveys were also usually restricted to studies of bright stars that were slow rotators. A search for the presence of magnetic fields in massive early B-type stars located in active sites of star formation, in the ρ Ophiuchus star-forming cloud and in the Trifid nebula, led to the detection of magnetic fields of several kG in two early B-type stars, the B1V star HD 164492Cb and the B2V star HD 147933 (Hubrig et al. 2014b, 2018; Pillitteri et al. 2018). While magnetic early B-type stars are frequently found close to the ZAMS, the position of magnetic O stars in the H-R diagram (Castro et al. 2015) indicates that spectropolarimetric observations of stars already evolved from the ZAMS bear a much larger potential to detect magnetic fields.

It is possible that since very young massive stars at the ZAMS are usually faint (Arias et al. 2016), this evolutionary stage of massive stars remains rather unexplored. On the other hand, a recent study of Holgado et al. (2020) of the position of more than 400 Galactic O-type stars in the spectroscopic Hertzsprung–Russell (sHR; Langer & Kudritzki 2014) diagram using high-quality spectroscopic observations gathered in different databases showed an apparent lack of stars with initial evolutionary masses in the range between ~30 and 70 $M_{\odot}$ near the ZAMS. The authors performed quantitative spectroscopy of Galactic O stars, based on different methods and stellar atmosphere codes and concluded that the lack of O-type stars detectable in optical wavelengths and with parameters compatible with the theoretical ZAMS is a real and robust empirical fact. This empirical evidence presents an important challenge for current theories related to the formation and evolution of massive stars.

A few years ago, Fossati et al. (2016) concluded that the fraction of massive magnetic stars decreases for larger evolutionary age. However, the age of magnetic stars determined with non-magnetic models is not accurate, meaning that stellar evolution models have to take the impact of magnetic fields into account. Furthermore, as discussed above, even for non-magnetic stars, an apparent lack of massive stars near the ZAMS was detected by Holgado et al. (2020), indicating a persisting discrepancy between models and observations. Currently, any study of the impact of strong magnetic fields in massive O stars on their evolution remains inconclusive due to difficulties in the interpretations of the observations.

If magnetic fields were solely produced by mergers, there should be no magnetic star in close binaries. On the other hand, as two thirds of all Galactic O-type stars were reported to be members of close binary systems (Sana et al. 2012), the evolution of a large fraction of massive stars is affected by environmental effects. In dense star clusters, close multibody interactions may cause single and binary stars to be scattered out of the star cluster (e.g., Leonard & Duncan 1990). Previous calculations of the kinematic characteristics of the three magnetic Of?p stars HD 108, HD 148937, and HD 191612 (Hubrig et al. 2011c) using Hipparcos data showed that they are moving at rather high space velocities, suggesting that they can be considered as candidate runaway stars (Hubrig et al. 2011c).

In spite of the fact that the binary fraction of massive stars is very high, only a handful of magnetic close binaries with P_{orb} below 20–30 days are known among massive O-type and early B-type stars. This is unexpected and is probably related to the magnetic field origin. Ferrario et al. (2009) suggested that magnetic Ap/Bp stars are products of a merger of two lower mass protostars. In this scenario, the binaries that we observe now were triple systems earlier in their history. Among the very few known SB2 systems with massive magnetic Bp components (e.g., Landstreet et al. 2017), the primary in the SB2 system HD 149277 exhibits the strongest magnetic field and the strongest subsynchronous rotation (González et al. 2018). Only one massive binary with magnetic field detections in both components, the eccentric B2V/B3V SB2 system ϵ Lupi with $P_{orb} = 4.56$ days, is known (Shultz et al. 2015). The presence of a field in this system was originally reported by Hubrig et al. (2009, 2011b) who used FORS 1 spectropolarimetric observations, but were not able to detect the secondary magnetic component due to the insufficient low spectral resolution of this instrument. It is of interest that this SB2 system has a distant tertiary companion on an orbit with $P_{orb} \approx 740$ yr (Zirm 2007). Among the lower-mass counterparts, only two close binary systems with Ap magnetic components, HD 98088 (e.g., Babcock 1958; Abt et al. 1968) and HD 161701 (Hubrig et al. 2014a) are known. Importantly, only for these two systems was the structure of their magnetic fields and the impact of the presence of a companion on this structure sufficiently studied. Further systematic searches for magnetic fields in binary systems should be conducted to properly evaluate theoretical models of the origin of the magnetic fields.

The occurrence of magnetic fields in massive stars can be much higher than 10%. Multiwavelength observations of these stars frequently show cyclical line-profile structures, identified as DACs and their presence is indicative of the existence of large-scale structures in the dense, supersonic outflows (Mullan 1984). As discussed in Section 5.2.1, DACs are understood to be induced by irregularities on the stellar surface. These irregularities relate either to temperature spots or non-radial pulsations (Cranmer & Owocki 1996). Action of a subsurface convection zone (Cantiello et al. 2009) would be the most likely driving mechanism that generates magnetic spots as the source of these perturbations. According to Cantiello & Braithwaite (2011), flux variations can be related to bright spots that arise due to the lower density of the gas in surface regions with higher magnetic pressure.

From Kepler and Kepler K2 photometry, Balona (2013, 2016) concluded that about 40% of A-type stars and 30%–50% of late O- and B-type stars show rotational modulation due to the presence of star spots. Pulsating stars, eclipsing binaries and ellipsoidal variables of high amplitude were avoided in these studies. Since it is difficult to conceive of starspot formation without the presence of a magnetic field, these results would indicate that a significant fraction of the magnetic massive stars remains uncovered due to a more complex magnetic field topology related to the presence of a subsurface convection zone. This zone may be the source of a global magnetic field, winding up toroidally with stochastic buoyancy breakouts at the surface and causing corotating magnetic bright spots at the surface of the star (Cantiello & Braithwaite 2011). To improve the statistics on the occurrence of magnetic fields in massive stars, spectropolarimetric time series are necessary for stars showing the presence of temperature spots. Such observations can be preferably carried out with spectropolarimeters installed at telescopes with large collecting areas permitting to greatly increase the sensitivity of the magnetic field measurements.

The fundamental parameters of a sample consisting of 56 B0–B5 magnetic stars were studied by Shultz et al. (2019). This sample was neither volume limited nor brightness limited. The study indicated that the range of surface magnetic field strengths is essentially constant with stellar mass and that the unsigned surface magnetic flux increases with mass. Both the surface magnetic dipole strength and the total magnetic flux were reported to decrease with stellar age, with the rate of flux decay apparently increasing with stellar mass.

No theoretical simulations assuming realistic MHD assumptions were carried out yet with respect to evolutionary changes of the surface magnetic field. Most theoretical work in the last ten years concentrated on the significance of the effects of magnetic braking and magnetic confinement. Future self-consistent simulations of stellar magnetism including wind mass loss due to wind–magnetic field interaction and realistic assumptions on the internal magnetic fields, internal differential rotation, chemical mixing, etc. are needed to get a more detailed insight into the evolution of magnetic massive O-type and early B-type stars.

References

Abt, H. A., Conti, P. S., Deutsch, A. J., & Wallerstein, G. 1968, ApJ, 153, 177

Appenzeller, I., Fricke, K., Fürtig, W., et al. 1998, Msngr, 94, 1

Arias, J. I., Walborn, N. R., Simón Díaz, S., et al. 2016, AJ, 152, 31

Auvergne, M., Bodin, P., Boisnard, L., et al. 2009, A&A, 506, 411

Babcock, H. W. 1958, ApJS, 3, 141

Babel, J., & Montmerle, T. 1997a, ApJL, 485, L29

Babel, J., & Montmerle, T. 1997b, A&A, 323, 121

Balona, L. A. 2013, MNRAS, 431, 2240

Balona, L. A. 2016, MNRAS, 457, 3724

Barron, J., Wade, G. A., Bowman, D. M., et al. 2020, in Proc. Polish Astronomical Soc. 11, Stellar Magnetism: A Workshop in Honour of the Career and Contributions of John D. Landstreet, ed. G. Wade, E. Alecian, D. Bohlender, & A. Sigut, 226

Blomme, R., Mahy, L., Catala, C., et al. 2011, A&A, 533, A4
Borucki, W. J., Koch, D., Basri, G., et al. 2010, Sci, 327, 977
Bouret, J. C., Donati, J. F., Martins, F., et al. 2008, MNRAS, 389, 75
Bowman, D. M., Burssens, S., Simón-Díaz, S., et al. 2020, A&A, 640, A36
Braithwaite, J., & Cantiello, M. 2013, MNRAS, 428, 2789
Braithwaite, J., & Spruit, H. C. 2004, Natur, 431, 819
Briquet, M., Aerts, C., Lüftinger, T., et al. 2004, A&A, 413, 273
Brott, I., Evans, C. J., Hunter, I., et al. 2011, A&A, 530, A116
Burssens, S., Simón-Díaz, S., Bowman, D. M., et al. 2020, A&A, 639, A81
Cantiello, M., & Braithwaite, J. 2011, A&A, 534, A140
Cantiello, M., Langer, N., Brott, I., et al. 2009, A&A, 499, 279
Castro, N., Fossati, L., Hubrig, S., et al. 2015, A&A, 581, A81
Castro, N., Fossati, L., Hubrig, S., et al. 2017, A&A, 597, L6
Charbonneau, P., & MacGregor, K. B. 2001, ApJ, 559, 1094
Cranmer, S. R., & Owocki, S. P. 1996, ApJ, 462, 469
Das, B., Chandra, P., Shultz, M. E., & Wade, G. A. 2019, MNRAS, 489, L102
David-Uraz, A., Erba, C., Petit, V., et al. 2019, MNRAS, 483, 2814
Dekker, H., D'Odorico, S., Kaufer, A., Delabre, B., & Kotzlowski, H. 2000, Proc. SPIE, 4008, 534
Donati, J. F., Babel, J., Harries, T. J., et al. 2002, MNRAS, 333, 55
Donati, J. F., Howarth, I. D., Bouret, J. C., et al. 2006, MNRAS, 365, L6
Dziembowski, W. A., Moskalik, P., & Pamyatnykh, A. A. 1993, MNRAS, 265, 588
Eikenberry, S. S., Chojnowski, S. D., Wisniewski, J., et al. 2014, ApJL, 784, L30
Ferrario, L., Pringle, J. E., Tout, C. A., & Wickramasinghe, D. T. 2009, MNRAS, 400, L71
Fossati, L., Castro, N., Morel, T., et al. 2015, A&A, 574, A20
Fossati, L., Schneider, F. R. N., Castro, N., et al. 2016, A&A, 592, A84
Friend, D. B., & MacGregor, K. B. 1984, ApJ, 282, 591
Gaensler, B. M., McClure-Griffiths, N. M., Oey, M. S., et al. 2005, ApJL, 620, L95
González, J. F., Hubrig, S., Järvinen, S. P., & Schöller, M. 2018, MNRAS, 481, L30
González, J. F., Briquet, M., Przybilla, N., et al. 2019, A&A, 626, A94
Groote, D., & Hunger, K. 1997, A&A, 319, 250
Grunhut, J. H., Wade, G. A., Marcolino, W. L. F., et al. 2009, MNRAS, 400, L94
Grunhut, J. H., Rivinius, T., Wade, G. A., et al. 2012, MNRAS, 419, 1610
Grunhut, J. H., Wade, G. A., Leutenegger, M., et al. 2013, MNRAS, 428, 1686
Grunhut, J. H., Wade, G. A., Neiner, C., et al. 2017, MNRAS, 465, 2432
Heger, A., Woosley, S. E., & Spruit, H. C. 2005, ApJ, 626, 350
Holgado, G., Simón-Díaz, S., Haemmerlé, L., et al. 2020, A&A, 638, A157
Howarth, I. D., & van Leeuwen, F. 2019, MNRAS, 484, 5350
Howarth, I. D., Prinja, R. K., & Willis, A. J. 1984, MNRAS, 208, 525
Howarth, I. D., Goss, K. J. F., Stevens, I. R., Chaplin, W. J., & Elsworth, Y. 2014, MNRAS, 440, 1674
Howell, S. B., Sobeck, C., Haas, M., et al. 2014, PASP, 126, 398
Hubrig, S., North, P., & Schöller, M. 2007, AN, 328, 475
Hubrig, S., Schöller, M., Schnerr, R. S., et al. 2008, A&A, 490, 793
Hubrig, S., Briquet, M., De Cat, P., et al. 2009, AN, 330, 317
Hubrig, S., Ilyin, I., Briquet, M., et al. 2011a, A&A, 531, L20

Hubrig, S., Ilyin, I., Schöller, M., et al. 2011b, ApJL, 726, L5
Hubrig, S., Kharchenko, N. V., & Schöller, M. 2011c, AN, 332, 65
Hubrig, S., Oskinova, L. M., & Schöller, M. 2011d, AN, 332, 147
Hubrig, S., Schöller, M., Kharchenko, N. V., et al. 2011e, A&A, 528, A151
Hubrig, S., Schöller, M., Ilyin, I., et al. 2013, A&A, 551, A33
Hubrig, S., Carroll, T. A., Gonzalez, J. F., et al. 2014a, MNRAS, 440, L6
Hubrig, S., Fossati, L., Carroll, T. A., et al. 2014b, A&A, 564, L10
Hubrig, S., Schöller, M., Fossati, L., et al. 2015a, A&A, 578, L3
Hubrig, S., Schöller, M., Kholtygin, A. F., et al. 2015b, MNRAS, 447, 1885
Hubrig, S., Kholtygin, A., Ilyin, I., Schöller, M., & Oskinova, L. M. 2016, ApJ, 822, 104
Hubrig, S., Kholtygin, A. F., Schöller, M., & Ilyin, I. 2017a, MNRAS, 467, L81
Hubrig, S., Mikulášek, Z., Kholtygin, A. F., et al. 2017b, MNRAS, 472, 400
Hubrig, S., Przybilla, N., Korhonen, H., et al. 2017c, MNRAS, 471, 1543
Hubrig, S., Schöller, M., Järvinen, S. P., et al. 2018, AN, 339, 72
Hubrig, S., Järvinen, S. P., Schöller, M., & González, J. F. 2019a, in ASP Conf. Ser. 519, Radiative Signatures from the Cosmos, ed. K. Werner, C. Stehle, T. Rauch, & T. Lanz (San Francisco, CA: ASP), 193
Hubrig, S., Küker, M., Järvinen, S. P., et al. 2019b, MNRAS, 484, 4495
Hubrig, S., Järvinen, S. P., Schöller, M., & Hummel, C. A. 2020, MNRAS, 491, 281
Hunger, K., & Groote, D. 1999, A&A, 351, 554
Järvinen, S. P., Hubrig, S., Ilyin, I., Schöller, M., & Briquet, M. 2017, MNRAS, 464, L85
Järvinen, S. P., Hubrig, S., Schöller, M., et al. 2021, MNRAS, 501, 4534
Küker, M. 2017, AN, 338, 868
Kambe, E., Ando, H., & Hirata, R. 1993, A&A, 273, 435
Kambe, E., Hirata, R., Ando, H., et al. 1997, ApJ, 481, 406
Karitskaya, E. A., Bochkarev, N. G., Hubrig, S., et al. 2009, in IAU Symp. 259, Cosmic Magnetic Fields: From Planets, to Stars and Galaxies, ed. K. G. Strassmeier, A. G. Kosovichev, & J. E. Beckman (Cambridge: Cambridge Univ. Press), 137
Koch, D. G., Borucki, W. J., Basri, G., et al. 2010, ApJL, 713, L79
Kurtz, D. W., Hubrig, S., Järvinen, S. P., & Schöller, M. 2020, RNAAS, 4, 157
Landstreet, J. D., Kochukhov, O., Alecian, E., et al. 2017, A&A, 601, A129
Langer, N., & Kudritzki, R. P. 2014, A&A, 564, A52
Leonard, P. J. T., & Duncan, M. J. 1990, AJ, 99, 608
Leto, P., Trigilio, C., Buemi, C. S., Leone, F., & Umana, G. 2012, MNRAS, 423, 1766
Maeder, A., & Meynet, G. 2005, A&A, 440, 1041
Martins, F., Donati, J. F., Marcolino, W. L. F., et al. 2010, MNRAS, 407, 1423
Massa, D., & Prinja, R. K. 2015, ApJ, 809, 12
Meynet, G., Eggenberger, P., & Maeder, A. 2011, A&A, 525, L11
Miglio, A., Montalbán, J., & Dupret, M.-A. 2007, MNRAS, 375, L21
Mikulášek, Z., Krtička, J., Zverko, J., Žižňovský, J., & Janík, J. 2007, in ASP Conf. Ser. 361, Active OB-Stars: Laboratories for Stellar and Circumstellar Physics, ed. A. T. Okazaki, S. P. Owocki, & S. Stefl (San Francisco, CA: ASP), 466
Mikulášek, Z., Krtička, J., Henry, G. W., et al. 2008, A&A, 485, 585
Mikulášek, Z., Krtička, J., Henry, G. W., et al. 2010, A&A, 511, L7
Mikulášek, Z., Krtička, J., Henry, G. W., et al. 2011, A&A, 534, L5

Morel, T., Castro, N., Fossati, L., et al. 2015, in IAU Symp. 307, New Windows on Massive Stars, ed. G. Meynet, C. Georgy, J. Groh, & P. Stee (Cambridge: Cambridge Univ. Press), 342
Moss, D. 2001, in ASP Conf. Ser. 248, Magnetic Fields Across the Hertzsprung–Russell Diagram, ed. G. Mathys, S. K. Solanki, & D. T. Wickramasinghe (San Francisco, CA: ASP), 305
Mullan, D. J. 1984, ApJ, 283, 303
Nazé, Y., Walborn, N. R., Rauw, G., et al. 2008, AJ, 135, 1946
Nazé, Y., Bagnulo, S., Petit, V., et al. 2012, MNRAS, 423, 3413
Nazé, Y., Petit, V., Rinbrand, M., et al. 2014, ApJS, 215, 10
Nazé, Y., Barbá, R., Bagnulo, S., et al. 2016, A&A, 596, A44
Neiner, C., Landstreet, J. D., Alecian, E., et al. 2012, A&A, 546, A44
Nichols, J. S., Nazé, Y., Huenemoerder, D. P., et al. 2021, ApJ, 906, 89
Nomoto, K., Maeda, K., Tanaka, M., & Suzuki, T. 2011, Ap&SS, 336, 129
Oskinova, L. M., Nazé, Y., Todt, H., et al. 2014, NatCo, 5, 4024
Pedersen, M. G., Chowdhury, S., Johnston, C., et al. 2019, ApJL, 872, L9
Perryman, M. A. C., Lindegren, L., Kovalevsky, J., et al. 1997, A&A, 500, 501
Petit, V., Wade, G. A., Drissen, L., Montmerle, T., & Alecian, E. 2008, MNRAS, 387, L23
Petit, V., Owocki, S. P., Wade, G. A., et al. 2013, MNRAS, 429, 398
Petit, V., Cohen, D. H., Wade, G. A., et al. 2015, MNRAS, 453, 3288
Pillitteri, I., Fossati, L., Castro Rodriguez, N., Oskinova, L., & Wolk, S. J. 2018, A&A, 610, L3
Pollmann, E. 2012, IBVS, 6034, 1
Queloz, D., Mayor, M., Weber, L., et al. 2000, A&A, 354, 99
Ramiaramanantsoa, T., Moffat, A. F. J., Chené, A.-N., et al. 2014, MNRAS, 441, 910
Ramiaramanantsoa, T., Moffat, A. F. J., Harmon, R., et al. 2018, MNRAS, 473, 5532
Reid, A. H. N., Bolton, C. T., Crowe, R. A., et al. 1993, ApJ, 417, 320
Ricker, G. R., Winn, J. N., Vanderspek, R., et al. 2015, JATIS, 1, 014003
Rivinius, T., Štefl, S., Townsend, R. H. D., & Baade, D. 2008, A&A, 482, 255
Rivinius, T., Townsend, R. H. D., Kochukhov, O., et al. 2013, MNRAS, 429, 177
Sana, H., de Mink, S. E., de Koter, A., et al. 2012, Sci, 337, 444
Schöller, M., Hubrig, S., Fossati, L., et al. 2017, A&A, 599, A66
Shenar, T., Oskinova, L. M., Järvinen, S. P., et al. 2017, A&A, 606, A91
Shultz, M., & Wade, G. A. 2017, MNRAS, 468, 3985
Shultz, M., Wade, G. A., Alecian, E., & BinaMIcS Collaboration, 2015, MNRAS, 454, L1
Shultz, M. E., Wade, G. A., Rivinius, T., et al. 2018, MNRAS, 475, 5144
Shultz, M. E., Wade, G. A., Rivinius, T., et al. 2019, MNRAS, 490, 274
Sikora, J., Wade, G. A., Bohlender, D. A., et al. 2015, MNRAS, 451, 1928
Smith, N., Li, W., Filippenko, A. V., & Chornock, R. 2011, MNRAS, 412, 1522
Stahl, O. 1998, in ESO Astrophysics Symp., Cyclical Variability in Stellar Winds, ed. L. Kaper, & A. W. Fullerton (Berlin: Springer), 246
Stello, D., Cantiello, M., Fuller, J., Garcia, R. A., & Huber, D. 2016, PASA, 33, e011
Sundqvist, J. O., ud-Doula, A., Owocki, S. P., et al. 2012, MNRAS, 423, L21
Tetzlaff, N., Neuhäuser, R., Hohle, M. M., & Maciejewski, G. 2010, MNRAS, 402, 2369
Townsend, R. H. D., & Owocki, S. P. 2005, MNRAS, 357, 251
Townsend, R. H. D., Owocki, S. P., & Groote, D. 2005, ApJL, 630, L81
Townsend, R. H. D., Oksala, M. E., Cohen, D. H., Owocki, S. P., & ud-Doula, A. 2010, ApJL, 714, L318
ud-Doula, A., & Nazé, Y. 2016, AdSpR, 58, 680

ud-Doula, A., & Owocki, S. P. 2002, ApJ, 576, 413
ud-Doula, A., Owocki, S. P., & Townsend, R. H. D. 2008, MNRAS, 385, 97
ud-Doula, A., Owocki, S. P., & Townsend, R. H. D. 2009, MNRAS, 392, 1022
Uzdensky, D. A., & MacFadyen, A. I. 2006, ApJ, 647, 1192
Wade, G. A., Alecian, E., Bohlender, D. A., et al. 2011, in IAU Symp. 272, Active OB Stars: Structure, Evolution, Mass Loss, and Critical Limits, ed. C. Neiner, G. Wade, G. Meynet, & G. Peters (Cambridge: Cambridge Univ. Press), 118
Wade, G. A., Grunhut, J., Gräfener, G., et al. 2012a, MNRAS, 419, 2459
Wade, G. A., Maíz Apellániz, J., Martins, F., et al. 2012b, MNRAS, 425, 1278
Walborn, N. R., Sota, A., Maíz Apellániz, J., et al. 2010, ApJL, 711, L143
Walborn, N. R. 1972, ApJL, 176, L119
Walborn, N. R. 1973, AJ, 78, 1067
Walker, G. A. H., Kuschnig, R., Matthews, J. M., et al. 2005, ApJL, 623, L145
Weber, E. J., & Davis, L. J. 1967, ApJ, 148, 217
Wickramasinghe, D. T., Tout, C. A., & Ferrario, L. 2014, MNRAS, 437, 675
Wilson, J. C., Henderson, C. P., Herter, T. L., et al. 2004, Proc. SPIE, 5492, 1295
Wisniewski, J. P., Chojnowski, S. D., Davenport, J. R. A., et al. 2015, ApJL, 811, L26
Yoon, S. C., Langer, N., & Norman, C. 2006, A&A, 460, 199
Ziegler, U. 2008, CoPhC, 179, 227
Ziegler, U. 2011, NIRVANA: A Numerical Tool for Astrophysical Gas Dynamics, ascl:1101.006
Zirm, H. 2007, IAU Commission 26, Double Stars, 161, 1

Chapter 6

Herbig Ae/Be Stars

Models of magnetically driven accretion reproduce many observational properties of T Tauri stars. For the more massive Herbig Ae/Be stars, the corresponding picture has been questioned, in part driven by the fact that their magnetic fields are typically one order of magnitude weaker. Indeed, the search for magnetic fields in Herbig Ae/Be stars is quite time consuming, limited by the current potential to detect weak magnetic fields. Over the last two decades, magnetic fields were found in about twenty objects and for only two Herbig Ae/Be stars was the magnetic field geometry constrained. Several studies were undertaken to investigate the time dependence of spectroscopic tracers of magnetospheric accretion. It appears that while there is proof that magnetospheric accretion is present in Herbig Ae stars, there is less evidence for the Herbig Be stars.

6.1 General Background

Studies of magnetic fields in stars at early evolutionary stages, before they arrive on the main sequence, are of special interest to get an insight into the magnetic field origin. It is generally accepted that magnetic fields are important ingredients of the star formation process (e.g., McKee & Ostriker 2007; Krumholz & Federrath 2019; McKee et al. 2020) and are already present in stars in the pre-main-sequence (PMS) phase. Herbig Ae/Be stars are optically visible PMS stars whose masses range from about 2 to 10 $M_{\odot}$ and are considered as interface targets between the low-mass and high-mass star formation. They were first identified as a group in the work of Herbig (1960) who studied Ae/Be stars associated with a nebulosity and selected a sample of 26 stars based on three criteria: (a) The star has a spectral type A or B with emission lines, (b) it is located in an obscured region, and (c) the star illuminates a bright nebulosity in its immediate vicinity. Later on, catalogues of Herbig Ae/Be stars were published by The et al. (1994) and Vieira et al. (2003). While low-mass star formation is relatively well understood, high-mass star formation is still unclear. Simulations of the star formation process usually concentrate on the low-mass end

doi:10.1088/2514-3433/abefccch6

of the stellar mass spectrum because computations of the formation of more massive stars are more complicated, as massive stars form in more massive molecular clouds and radiative feedback needs to be taken fully into account (e.g., Krumholz & Matzner 2009).

The low-mass T Tauri stars are known as solar-mass PMS stars, identified by their unusual optical spectra, which include bright Hα emission lines and numerous permitted and forbidden atomic transitions. In contrast to classical T Tauri stars (cTTSs), weak-line T Tauri stars (wTTSs) lack Hα emission in excess of 10 Å in equivalent widths (Herbig & Bell 1988). For cTTSs, the magnetic field and its geometry determine how and where the disk couples to the star. These stars are thought to be important in launching bi-polar outflows. According to Matt & Pudritz (2005), magnetic fields coupled to winds can be responsible for removing angular momentum from cTTS. Their model presents a synthesis of many results from the literature on disk winds, stellar winds, and funnel accretion flows. Furthermore, magnetic processes are responsible for the high energy radiation of young stars, which in turn may be critical for the annealing and melting of some refractory grain species (Shu et al. 2001), for heating intermediate-depth portions of disks around young stars (Najita 2004) and for mass loss from planets forming in those disks (Penz et al. 2008).

While there is general consensus that the magnetic fields of cTTS are generated in dynamo processes (Rosner et al. 1985) and that the magnetic fields of cTTS rule the accretion process (magnetospheric accretion; e.g., Hartmann et al. 1994; Muzerolle et al. 1998b), both the origin of the fields and the way they couple to the circumstellar material are not clear in the case of the Herbig Ae/Be stars. Observations of the disk properties of Herbig Ae/Be stars suggest a close parallel to T Tauri stars, revealing the same size range of the disks and similar optical surface brightness and structure. It is quite possible that magnetic fields play a crucial role in controlling accretion onto, and winds from, Herbig Ae stars, similar to the magnetospheric accretion observed in T Tauri stars. Notably, Vink et al. (2002) presented results of Hα spectropolarimetric observations of a sample of 23 Herbig Ae/Be stars and pointed out the possibility of the existence of a physical transition region in the H-R diagram from magnetospheric accretion at spectral type A, similar to that of cTTS, to disk accretion at spectral type B. The main difference between these scenarios is that in the former case the stellar magnetic field truncates the accretion disk at a few stellar radii and gas accretes along magnetic channels from the protoplanetary disk to the star, while in the latter case the accretion flow is not disrupted by the field. Cauley & Johns-Krull (2014) studied the He I λ10830 morphology in a sample of 56 Herbig Ae/Be stars. They concluded that while Herbig Ae stars show evidence for magnetospheric accretion, Herbig Be stars do not accrete material from their inner disks in the same manner as T Tauri stars. The structure of the inner disk of Herbig Be stars is, however, not well understood. The disks of several Herbig Be stars have inner radii that are smaller than predicted by models of irradiated disks with optically thin holes (e.g., Vural et al. 2014).

Although magnetospheric accretion Balmer and sodium line profiles in the Herbig Ae star UX Ori have been confirmed a long time ago by Muzerolle et al. (2004), current theories are not able to present a consistent scenario of how the magnetic fields in Herbig Ae/Be stars are generated and how these fields interact with the circumstellar environment, consisting of a combination of disk, wind, accretion and jets. On the other hand, understanding the interaction between the central stars, their magnetic fields and their protoplanetary disks is crucial for reconstructing the solar system's history, and to account for the diversity of exo-planetary systems. Protoplanetary disks, which are a natural outcome of the star formation process, are believed to be the environments where planets are formed. A considerable number of disks observed at high spatial resolution in dust continuum emission reveal disk misalignments and a multitude of substructures like dark gaps, bright rings, spiral arms, vortices, etc., which are indicative of planet formation. It is possible that the circularly-symmetric gaps and rings are caused by embedded planets, while spirals could be induced by external perturbers. Current multiwavelength observations of the protoplanetary disks of several Herbig Ae/Be stars, e.g., AB Aur (Boccaletti et al. 2020), HD 97048 (Pinte et al. 2019), HD 100453 (Nealon et al. 2020), HD 100546 (Pineda et al. 2019), HD 135344B (Stolker et al. 2017), HD 139614 (Muro-Arena et al. 2020), HD 163296 (Pinte et al. 2020), and HD 169142 (Bertrang et al. 2020), show many intriguing structures, both in scattered light and thermal IR emission, which are assumed to be related to planet formation processes and planet–disk interactions. In Figure 6.1 the scattered light images of the Herbig Ae star HD 139614 obtained by Muro-Arena et al. (2020) using the VLT/SPHERE instrument at four different epochs, three with a coronagraph and one without, are presented. In these images, the outer disk shows a striking azimuthal asymmetry with a brighter region located in the west/north-west. According to the authors, the observed broad shadows in the HD 139614 disk and the dust-depleted gap responsible for a dip in the SED can be explained by the presence of a planetary mass companion in the disk, located on an inclined orbit. Obviously, studies of the magnetic field topology using spectropolarimetry are extremely important because they should allow astronomers to improve the insight into how the magnetic fields in Herbig Ae/Be stars are generated and how they interact with their environment, including their impact on the planet formation processes and the planet–disk interaction.

6.2 Magnetic Field Studies of Herbig Ae/Be Stars

There are several reasons why the detection of magnetic fields in Herbig Ae/Be stars is difficult. The young intermediate-mass Herbig Ae stars as well as the more massive Herbig Be stars tend to be surrounded by large amounts of circumstellar gas and dust, making it hard to measure their magnetic fields. A number of spectral lines show circumstellar absorptions with complex profiles, and some lines appear in emission or exhibit P Cygni-type profiles. The optical spectra of Herbig Ae/Be stars contain fewer absorption lines compared to cooler stars, and the rotational velocities of A and B stars as a class tend to be higher. In addition, Herbig Ae stars are less common than T Tauri stars, since the IMF favours the production of lower mass stars, and, therefore, the sample of Herbig Ae stars includes rather faint objects. Several attempts in the past to measure magnetic fields in Herbig Ae/Be stars were

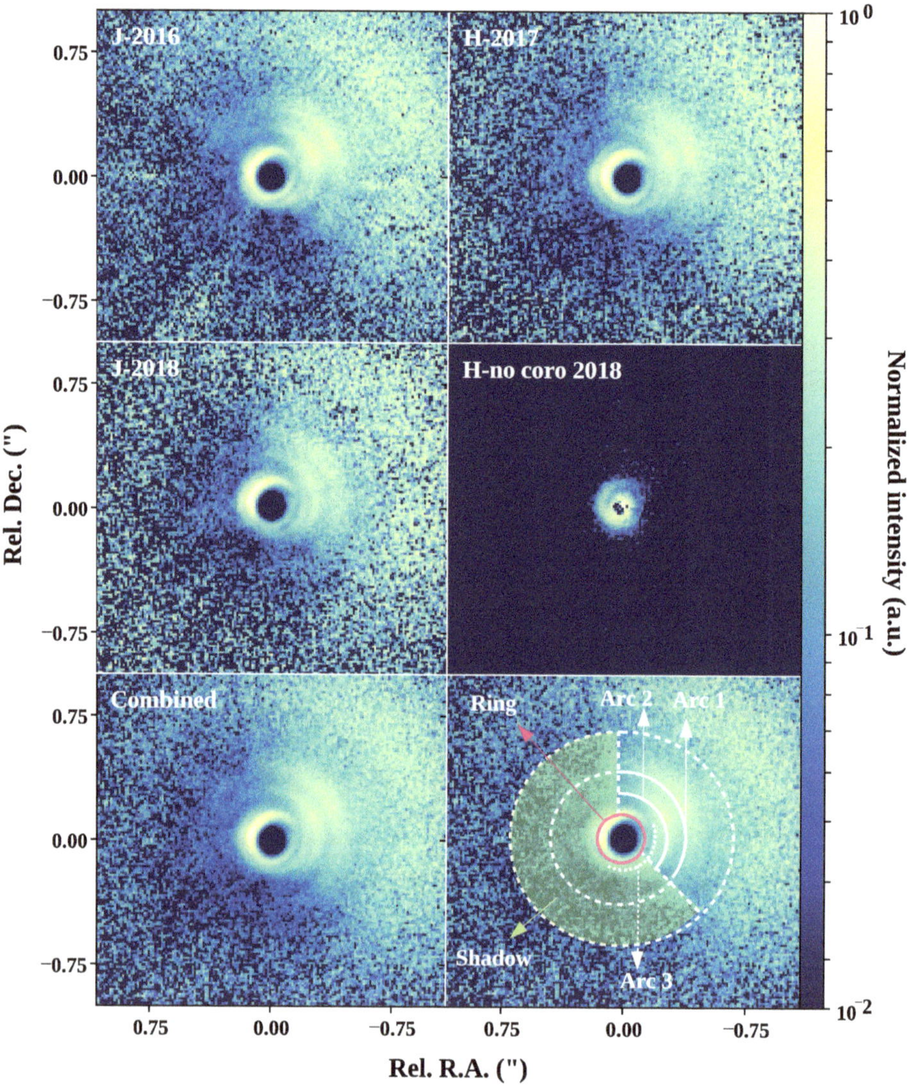

Figure 6.1. SPHERE/VLT scattered light observations of HD 139614, obtained at four different epochs (*top four panels*). The first three images (*top two and middle left panels*) were obtained with a coronagraphic mask. The next image (*middle right panel*) shows a non-coronagraphic image. The labels indicate the wavelength range and year of the observations. The bottom row shows the combined image of all coronagraphic images, with an annotated version pointing out the main features in the bottom right image. All images are normalised to their peak intensity after r^2-scaling, with the exception of the non-coronagraphic epoch, which is not r^2-scaled since this only accentuates the noise at large radii. Credit: Muro-Arena et al. (2020), reproduced with permission © ESO.

unsuccessful (e.g., Catala et al. 1993; Glagolevskij & Chountonov 2001). The first detection of a magnetic field was reported for the optically brightest Herbig Ae star HD 104237 viewed with the disk close to face-on by Donati et al. (1997). Using the Anglo-Australian Telescope with the UCL Echelle Spectrograph and a visitor polarimeter a marginal circular polarisation signature was observed in metallic lines, indicating a longitudinal magnetic field of about 50 G.

HD 104237 is a spectroscopic and visual binary system with a companion at a distance of 2.2 ± 0.7 mas (Garcia et al. 2013). The primary component in this systems is known to show δ Scuti-like pulsations (Böhm et al. 2004). Garcia et al. (2013) estimated the masses for both companions, $M_1 = 2.2 \pm 0.2\ M_\odot$ and $M_2 = 1.4 \pm 0.3\ M_\odot$, and a system inclination angle $i = 17^{\circ +12}_{\ -9}$. The orbital elements of the system, $P_{\rm orb} = 19.859$ days and $e = 0.665$, and the rotation period $P_{\rm rot} = 100 \pm 5$ h of the primary were reported by Böhm et al. (2004) and Böhm et al. (2006), respectively.

Advances in instrumentation over the last two decades have resulted in the first magnetic field detections in several Herbig Ae/Be stars. These detections were achieved using low- and high-resolution spectropolarimetry. The first spectropolarimetric observations of several Herbig Ae/Be stars using FORS 1 low-resolution circular polarisation observations showed evidence for large-scale organised magnetic fields reminiscent of the classical Ap/Bp stars, of which the magnetic Herbig Ae/Be stars were suggested to be the precursors (Hubrig et al. 2004). Using this instrument, magnetic field strengths of the order of tens to a few hundred Gauss were reported in a number of Herbig Ae/Be stars by Hubrig et al. (2004, 2006, 2007b) and Wade et al. (2005, 2007). Examples of detections of the mean longitudinal magnetic field using low-resolution FORS 1 spectropolarimetry in two Herbig Ae/Be stars, HD 101412 and HD 144668, are presented in Figures 6.2 and 6.3. A mean longitudinal magnetic field $\langle B_z \rangle$ of the order of −400 G was detected in HD 101412, whereas that for HD 144668 was of the order of −200 G. The first low-resolution polarimetric spectra with FORS 1 on the Very Large Telescope yielded however a non-detection of a magnetic field in HD 104237 (Wade et al. 2007). On the other hand, after high-resolution HARPSpol observations at the 3.6 m ESO

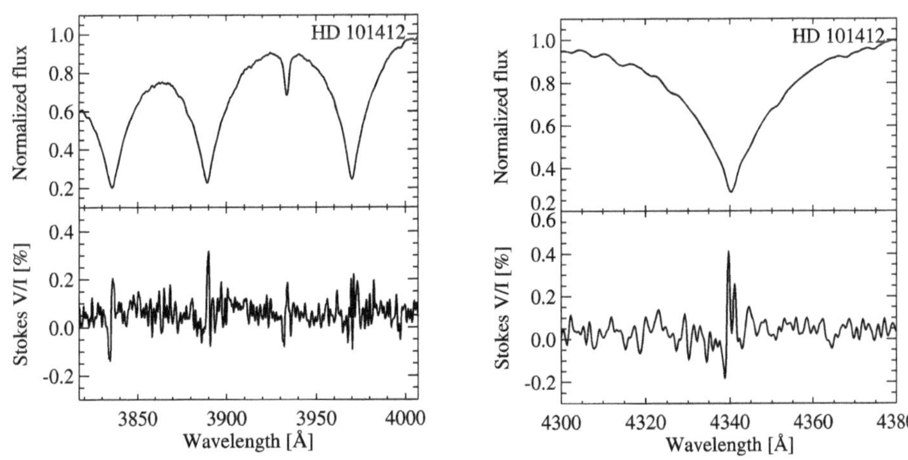

Figure 6.2. Stokes I and Stokes V spectra of the Herbig Ae/Be star HD 101412. *Left panel:* Zeeman features observed in H_9, H_8, Ca II H&K, and Hϵ profiles. *Right panel:* Stokes I and V spectra observed in the vicinity of the Hγ line. Credit: Hubrig et al. (2009), reproduced with permission © ESO.

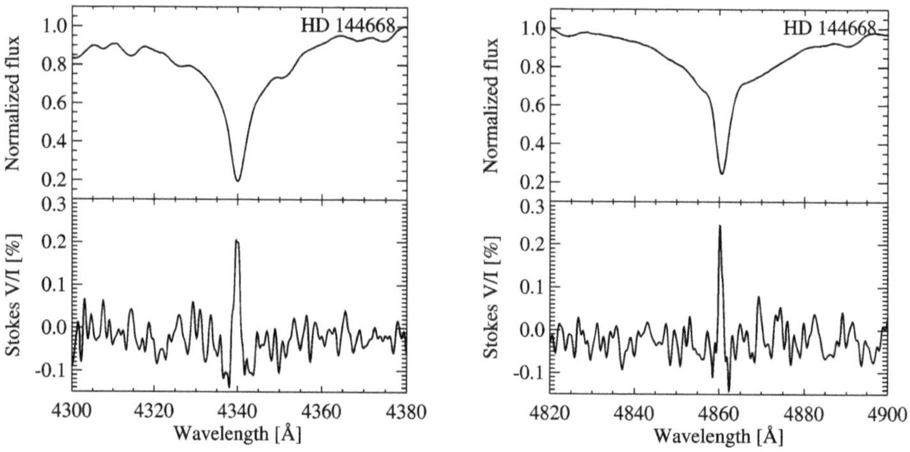

Figure 6.3. Stokes I and Stokes V spectra of the Herbig Ae/Be star HD 144668. *Left panel:* Zeeman features observed in the Hγ line. *Right panel:* Zeeman feature observed in the Hβ line. Credit: Hubrig et al. (2009), reproduced with permission © ESO.

telescope on La Silla became available, Hubrig et al. (2013) reported on the presence of a longitudinal field $\langle B_z \rangle = 63 \pm 15$ G, detected in the primary component of this system.

Owing to the presence of a rather strong magnetic field, the Herbig Ae star HD 101412 became one of the most extensively studied Herbig Ae stars in the last decades using observations with different instruments in the optical and the near-IR. After the detection of a magnetic field of the order of a few hundred Gauss using low-resolution spectropolarimetric observations, subsequent studies of this star showed that it possesses the strongest magnetic field ever measured in any Herbig Ae star, with a mean magnetic field modulus $\langle B \rangle$ of up to 3.5 kG. The inspection of UVES spectra retrieved from the ESO archive indicated distinct variability of line profiles belonging to different elements (Hubrig et al. 2010). In addition, the Zeeman doublet Fe II line at 6149.258 Å, which is the best diagnostic line to detect and measure surface magnetic fields in slowly rotating classical Ap stars with strong magnetic fields, appeared slightly resolved, suggesting the presence of a rather strong surface magnetic field. The study of this Zeeman doublet in Ap and Bp stars usually presents an excellent opportunity to determine in a straightforward, mostly approximation-free, model-independent way, and with particularly high precision the mean magnetic field modulus $\langle B \rangle$, that is, the average over the visible stellar hemisphere of the modulus of the magnetic vector, weighted by the local line intensity (see Section 2.6).

Currently, HD 101412 is the only Herbig Ae/Be star for which the rotational Doppler effect was found to be small in comparison to the magnetic splitting, with several spectral lines observed in unpolarised light resolved into magnetically split components (Hubrig et al. 2010). The Fe II λ6149.258 line exhibiting a resolved Zeeman doublet pattern is presented in Figure 6.4. A typical example for splitting in the spectrum of the bright A0p star HD 116458 with a magnetic field modulus of the

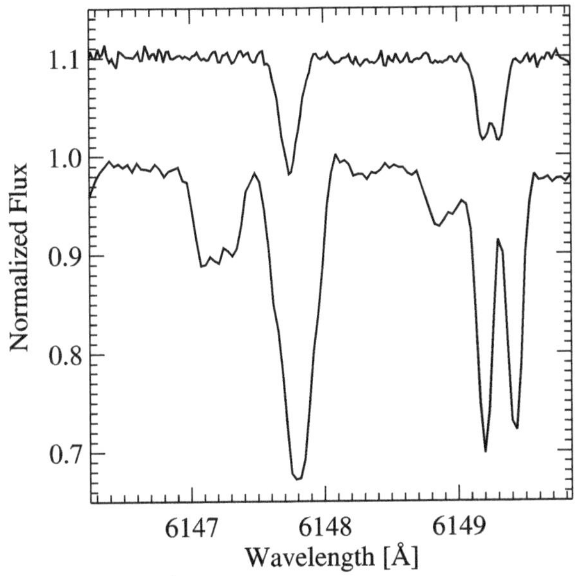

Figure 6.4. Portion of the spectra of HD 101412 (vertically shifted in intensity by 0.1 for clarity) and of the typical Ap star HD 116458 containing the Fe II λ6147.741 line with a Zeeman pseudo-quadruplet pattern and the λ6149.258 line with a Zeeman doublet pattern. Credit: Figure 2 from Hubrig et al. (2010).

order of 4.7 kG is shown for comparison in the lower panel of this figure. The behaviour of the split components of the Fe II λ6149.258 line in eight high signal-to-noise (S/N) UVES spectra and four lower quality HARPS spectra of HD 101412 is presented in Figure 6.5. The observed variable asymmetry of the split components is usually explained by a variable combination of Zeeman and Doppler effects across the stellar surface. In the same figure, the neighbouring line Fe II λ6147.741, which is a Zeeman pseudo-quadruplet, appears variable too and shows in some rotation phase a clearly triangular shape. Such a triangular shape indicates that the σ-components of this line are very weak at these epochs due to the predominance of a transversal field.

To constrain the rotation period and the geometry of the magnetic field of HD 101412, Hubrig et al. (2011a) acquired multi-epoch low-resolution polarimetric spectra with FORS 2. The analysis of the temporal variations of the measured longitudinal magnetic fields, making a frequency analysis and computing best-fit curves to these variations, permitted to ascertain the rotation/magnetic period (e.g., Hubrig et al. 2011a). The corresponding phase diagram for all available mean longitudinal magnetic field measurements using the whole spectrum and for those using only the hydrogen lines, including the best sinusoidal fit, is shown in Figure 6.6. The authors reported that the longitudinal magnetic field of HD 101412 varies with a rotation period of 42 days and has a mean $\langle \bar{B}_z \rangle = 9 \pm 18$ G and an amplitude of $A_{\langle B_z \rangle} = 465 \pm 27$ G.

As shown in Figure 6.6, the star exhibits a single-wave variation of the longitudinal magnetic field during the stellar rotation cycle. This variability is

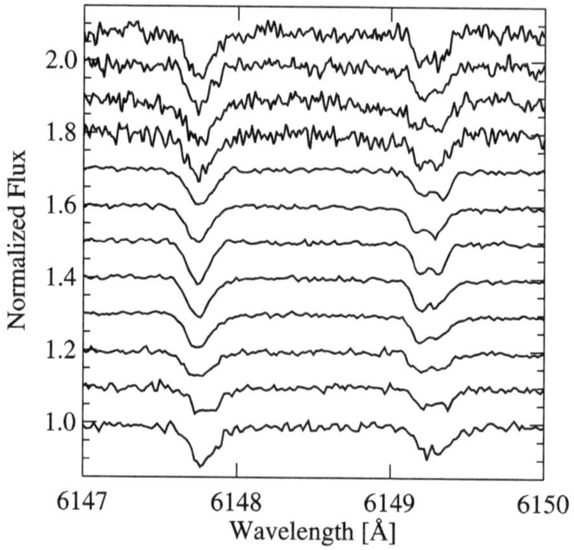

Figure 6.5. Line profile variations of the Zeeman doublet Fe II λ6149.258 in UVES and HARPS spectra of HD 101412 obtained at different epochs. Credit: Figure 3 from Hubrig et al. (2010).

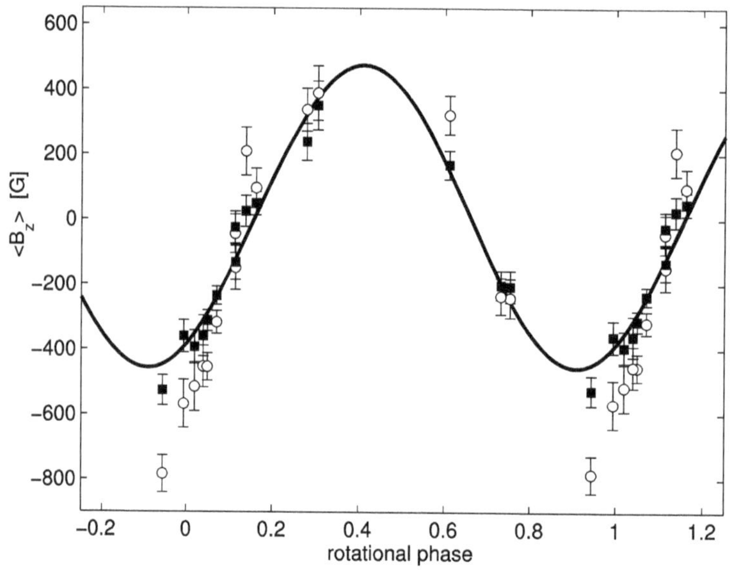

Figure 6.6. Phase diagram with the best sinusoidal fit for the FORS 2 longitudinal magnetic field measurements of HD 101412 using the entire spectrum (filled squares) or hydrogen lines (open circles). Credit: Hubrig et al. (2011a), reproduced with permission © ESO.

usually considered as evidence for a dominant dipolar contribution to the magnetic field topology. The study of the abundances of HD 101412 using UVES and HARPS spectra by Cowley et al. (2010) resulted in $v \sin i = 3 \pm 1$ km s^{-1}. The inclination angles of disks of Herbig Ae/Be stars (which are expected to be identical with the inclination angle of the stellar rotation axis) can be reliably derived only for resolved observations of disks. Fedele et al. (2008) used VLTI/MIDI observations to determine $i = 80 \pm 7°$. Using this value, $v_{eq} = 3 \pm 1$ km s^{-1} can be estimated. Assuming that HD 101412 is an oblique dipole rotator and following the formalism described in Section 2.3 (Equations (2.9) and (2.11)), one can determine the obliquity angle β. From the phase curve with $\langle B_z \rangle^{max}$=474 ± 32 G and $\langle B_z \rangle^{min}$=−456 ± 32 G follows $r = -0.962 \pm 0.075$, which for $i = 80 \pm 7°$ leads to a magnetic obliquity of $\beta = 84 \pm 13°$. The detection of a large-scale organised predominantly dipolar magnetic field on the surface of the young Herbig Ae star HD 101412 confirms the scenario that the accretion phenomenon in young stellar objects is magnetospheric accretion. Due to the presence of the rather strong magnetic field in the atmosphere of HD 101412, and a very low $v \sin i$-value of 3 ± 1 km s^{-1}, this star is one of the most suitable targets to study its environment, which includes the magnetosphere, the accretion disk, and the disk wind, altogether producing prominent emission features in the hydrogen line profiles (see Section 6.3).

The finding by Hubrig et al. (2011a) that the dipole axis in HD 101412 is located close to the stellar equatorial plane is very intriguing in view of the generally assumed magnetospheric accretion scenario that magnetic fields channel the accretion flows toward the stellar surface along the magnetic field lines. As was shown by Romanova et al. (2003), the topology of the channelled accretion critically depends on the tilt angle between the rotation axis and the magnetic axis. For large inclination angles β, many polar field lines would thread the inner region of the disk, while the closed lines cross the path of the disk matter, causing strong magnetic braking, which could explain the observed unusually long rotation period of HD 101412 of about 42 days.

The variations of equivalent width, radial velocity, line width, and line asymmetry of a sample of iron lines, the mean magnetic field modulus, and the mean longitudinal magnetic field as a function of the rotational phase are presented in Figure 6.7. Due to imperfect phase coverage, the interpretation of the observed variations of the iron line profiles was not straightforward, but it seemed that equivalent width and line asymmetry minima appear in the phase close to the surface field maximum of 3.5 kG and the positive extremum of the longitudinal field, while the presence of a secondary minimum is likely at the phase of the negative extremum of the longitudinal field. The line width is the largest at the phase close to the positive field extremum. The radial velocity variations are most difficult to interpret: it is possible that they show a double modulation, but it is not well-resolved in the measurements due to insufficient phase coverage (Hubrig et al. 2011a).

As of today, only about 20 Herbig stars were reported to host magnetic fields. Alecian et al. (2013) carried out a search for magnetic fields in 70 Herbig Ae/Be stars with masses in the range 1.5–16 $M_\odot$ and spectral types between F and B0 using high-resolution spectropolarimetric observations with ESPaDOnS at CFHT and Narval at TBL. The authors concluded that only very few of the studied Herbig Ae/Be stars possess longitudinal magnetic fields, indicating a detection rate of only 7%. However, the task of magnetic field measurements in Herbig Ae/Be stars is very challenging, as the work of Hubrig et al. (2015a) demonstrated, in which the authors

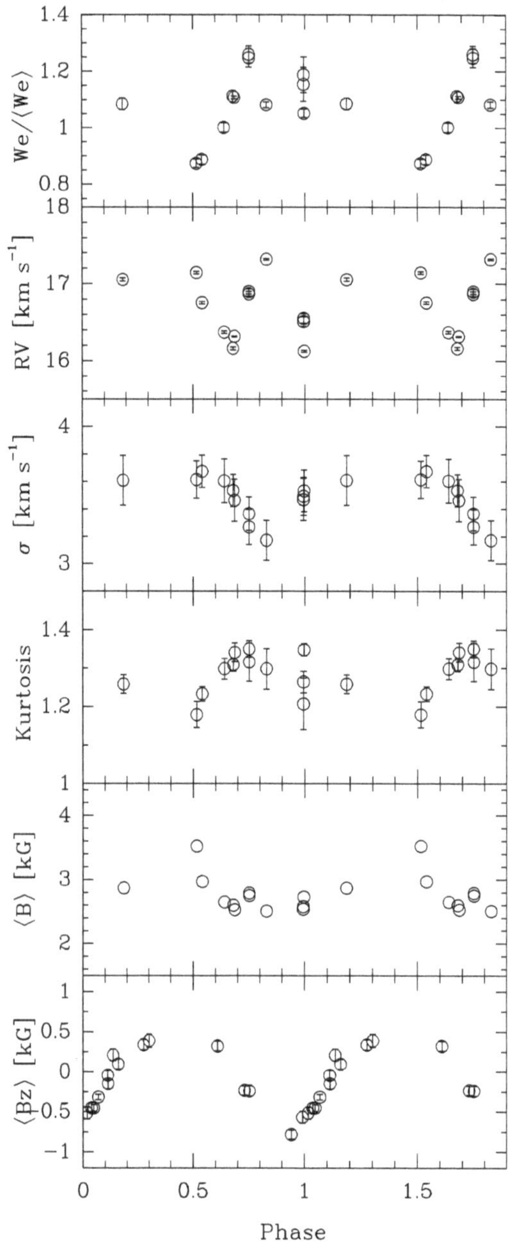

Figure 6.7. Variations of equivalent width, radial velocity, line width, line asymmetry, mean magnetic field modulus, and mean longitudinal magnetic field as a function of the rotational phase in HD 101412. Credit: Hubrig et al. (2011a), reproduced with permission © ESO.

compiled all magnetic field measurements reported in previous spectropolarimetric studies. This study suggested that the low detection rate of magnetic fields in Herbig Ae/Be stars can be plausibly explained by the limited sensitivity of the published measurements and by the weakness of the magnetic fields in these stars.

The obtained density distribution of the rms longitudinal magnetic field values for all late Herbig Be and Herbig Ae stars for which detections of a magnetic field were reported in the past is presented in Figure 6.8. It shows that only a few stars have magnetic fields stronger than 200 G, and half of the sample possesses magnetic fields of about 100 G and less. Consequently, the spectropolarimetric survey of magnetic fields in 70 Herbig Ae/Be stars by Alecian et al. (2013) cannot be considered as representative: the measurement accuracy in this study was worse than 200 G for 35% of the measurements, and for 32% of the measurements it was between 100 and 200 G. Clearly, only by improving the measurement accuracy it will become possible to realistically estimate the incidence rate of magnetic Herbig Ae/Be stars.

Among the Herbig Ae/Be stars with a measured magnetic field, the strongest longitudinal magnetic field for a Herbig Be star was measured in the system Z CMa using low-resolution FORS 1 spectropolarimetry (Szeifert et al. 2010). Z CMa is a binary system that consists of two young stars: a Herbig Be component Z CMa NW embedded in a dust cocoon and a less massive component Z CMa SE, which is classified as a FU Orionis type star. A giant parsec-size jet is associated with this binary system. In 2008, Z CMa showed the largest outburst reported during the almost 90 years of available observations. FORS 2 linear and circular spectropolarimetric observations of this system during the outburst were presented by Szeifert et al. (2010), who used linear polarisation spectra to study the geometry of the system and circular polarisation spectra to study the presence of a longitudinal magnetic field. The spectrum of the Z CMa system was found to be polarised by 2.6% in the continuum and emission line spectrum, with a position angle perpendicular to the jet. From the high level of polarisation the authors concluded that the

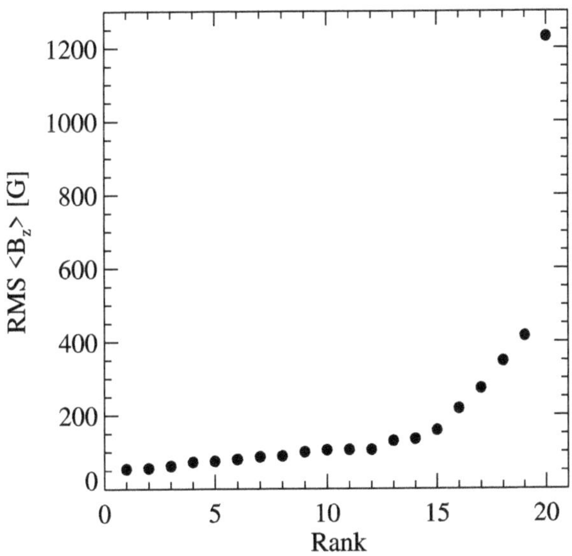

Figure 6.8. Density distribution of the rms longitudinal magnetic field values for the twenty late Herbig Be and Herbig Ae stars for which detections of a magnetic field were reported in the literature until 2015. Credit: Figure 3 from Hubrig et al. (2015a).

outburst is associated with the dust-embedded Herbig Be NW component. The deep absorption components of the Balmer lines in the velocity frame, which extended from zero velocity and reached a wind velocity of 700 km s^{-1}, together with the absence of a redshifted broad emission at similar velocities, indicated the presence of a bi-polar wind. No significant mean longitudinal magnetic field during the outburst was detected. However, a clear circular polarisation signal, in particular at the blueshifted side of the high number Balmer lines, was discovered in archival FORS 1 polarimetric spectra obtained in 2004. Using the low excitation Balmer lines, a lower field strength of -160 ± 84 G was detected, whereas for the high excitation Balmer lines the magnetic field was much stronger with $\langle B_z \rangle = -1231 \pm 164$ G. Since Z CMa is a binary star, it was unclear whether the magnetic field was detected in the dust-embedded Herbig Be NW component or in the FU Orionis type companion. Since the bolometric luminosity of Z CMa remained surprisingly constant during this outburst, Szeifert et al. (2010) concluded that either the geometry of the cavity through which the light escapes from the cocoon has opened a new path or that the screen of dust, which reflects the light toward the observer became more efficient, causing the observed increase of the visual brightness.

Zeeman signatures in the spectra of Herbig Ae/Be stars are generally very small, and increasing the S/N by increasing the exposure time in high-resolution spectropolarimetric observations is frequently limited by the shortness of the rotation periods of Herbig Ae/Be stars. Therefore, multi-line approaches as proposed by Semel (1989) are commonly used to increase the S/N. The main assumptions in the multi-line least squares deconvolution (LSD; Donati et al. 1997) technique are the application of the weak field approximation, i.e., the magnetic splitting of spectral lines is assumed to be smaller than their Doppler broadening, and that the local line profiles are self-similar and can be combined into an average profile. A software package to study weak magnetic fields using the multi-line singular value decomposition (SVD) method was introduced by Carroll et al. (2012; see Section 2.10). An example of a very weak Zeeman feature in the HARPSpol spectra of the Herbig Ae star PDS 2 detected using the SVD method is presented in Figure 6.9 (Hubrig et al. 2015a). No longitudinal

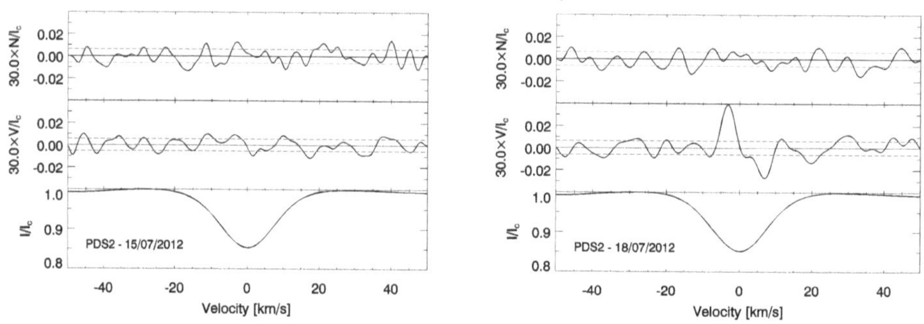

Figure 6.9. Stokes *I*, Stokes *V*, and diagnostic null *N* SVD profiles obtained for PDS 2 on two different nights. The Stokes *V* and diagnostic null *N* profiles were expanded by a factor of 30 and shifted upwards for better visibility. The red dashed lines indicate the standard deviations for the Stokes *V* and diagnostic null *N* spectra. Credit: Figure 2 from Hubrig et al. (2015a).

magnetic field was detected on the first epoch on 2012 July 15, where $\langle B_z \rangle = 5 \pm 5$ G was measured. A definite detection of a field, $\langle B_z \rangle = 33 \pm 5$ G, with a false alarm probability (FAP) smaller than 10^{-6}, was achieved on the second epoch three nights later on 2012 July 18. For both observations the null spectra appeared flat, indicating the absence of spurious polarisation.

The non-detection of the longitudinal magnetic field on the first epoch indicates that single snapshot observations are not sufficient to judge whether a Herbig Ae/Be star is magnetic or not. The longitudinal magnetic field is defined as the disk-integrated magnetic field component along the line of sight and therefore shows a strong dependence on the viewing angle of the observer, i.e., on the rotation angle of the star. The limitations set by the strong geometric dependence of the longitudinal magnetic field are usually overcome by repeating observations several times, so as to sample various rotation phases, hence various aspects of the magnetic field. Unfortunately, the rotation periods of Herbig Ae/Be stars are poorly known (e.g., Hubrig et al. 2011a, 2011b). The observed light variations in these stars are likely of stochastic nature, caused by obscuration of the circumstellar material and fluctuating disk accretion phenomena. Multi-epoch rotation-modulated longitudinal magnetic field measurements are frequently used to determine rotation periods, but such monitoring of a representative sample of Herbig Ae/Be stars with high-resolution spectropolarimetry has not been carried out yet.

New evidence for the presence of weak magnetic fields in Herbig Ae/Be stars was reported by Järvinen et al. (2019a), who used HARPSpol spectropolarimetric observations of additional five Herbig Ae/Be stars, HD 58647, HD 98922, HD 139614, HD 165133, and HD 190073. Applying the SVD method to these observations, the detected mean longitudinal magnetic fields in this sample ranged from about tens of Gauss to 200 G. The strongest longitudinal magnetic field of 209 G was detected in the late Herbig Be star HD 58647, whereas the weakest field of 17 G was measured in the Herbig Ae star HD 190073. A change of polarity was detected for the late-B stars HD 58647 and HD 98922. Kurosawa et al. (2016) studied the wind-launching region of the late Herbig Be star HD 58647 using high angular and high spectral resolution Very Large Telescope Interferometer (VLTI) Astronomical Multi-Beam Recombiner (AMBER) observations of the near-infrared hydrogen emission line Brγ. The authors reported that a model with a small magnetosphere and a disk wind with its inner radius located just outside of the magnetosphere can well reproduce the observed Brγ profile, wavelength-dependent visibilities, differential phases and closure phases, simultaneously. An example of modelling the interferometric observations of the Brγ line in the spectrum of this star is presented in Figure 6.10.

The SVD Stokes I and V profiles, and diagnostic null N profiles obtained for all five Herbig Ae/Be stars are shown in Figure 6.11. The results of Järvinen et al. (2019a) showed that magnetic fields in Herbig Ae/Be stars are by far weaker than fields measured in their lower mass counterparts, the T Tauri stars, usually possessing kG magnetic fields. The high S/N high-resolution HARPSpol observations presented in Figure 6.11 indicate that the intensity profiles in Herbig Ae/Be stars are frequently highly variable. The strong spectral variability of these stars is

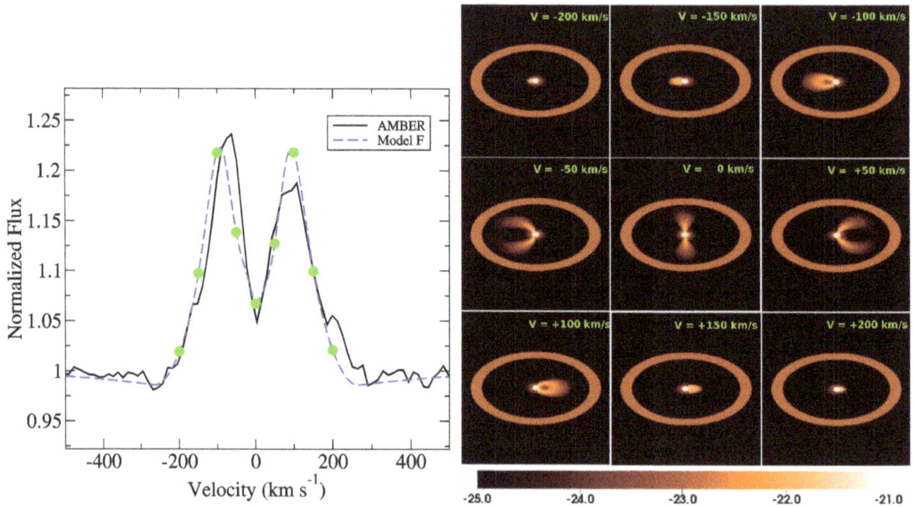

Figure 6.10. *Left panel:* the model Brγ profile (dash) computed at $i = 55°$ with the magnetosphere and disk wind is compared with the AMBER observation (solid) of HD 58647. The velocity channels at which the emission maps are computed (right panel) are indicated by filled circles. *Right panel:* the corresponding emission maps (in logarithmic scale and in arbitrary units) in nine different velocity channels. The wavelength/velocity dependent images are used to compute the interferometric quantities such as visibilities, differential phases and closure phases as a function of wavelength/velocity. Credit: Figure 7 from Kurosawa et al. (2016).

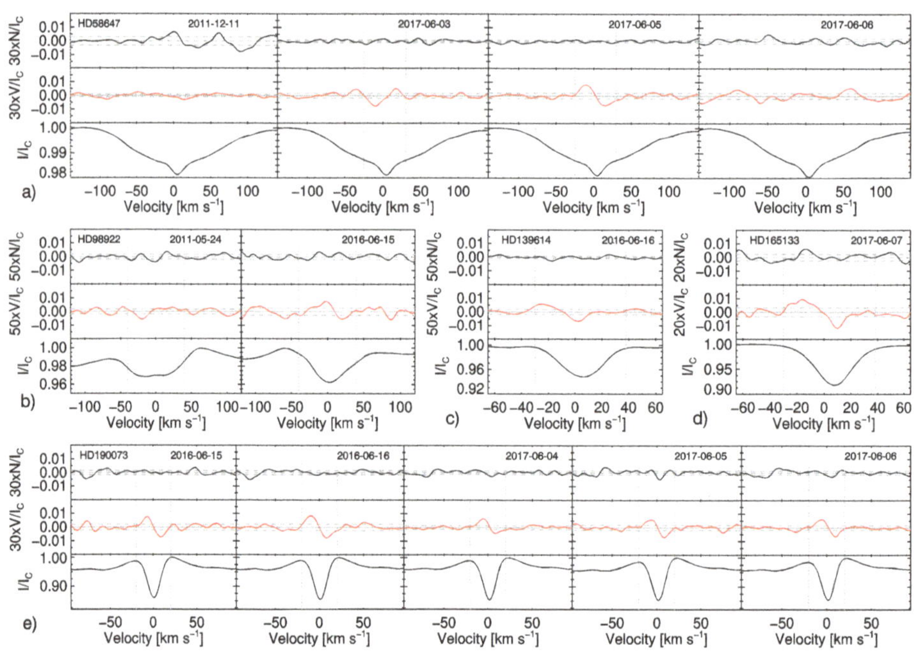

Figure 6.11. SVD Stokes *I* (*bottom*), Stokes *V* (*middle*), and diagnostic null (N) profiles (*top*) for the Herbig Ae/Be stars HD 58647, HD 98922, HD 139614, HD 165133, and HD 190073. Credit: Figure 1 from Järvinen et al. (2019a).

due to the complex interaction between the stellar magnetic field, the accretion disk, and the stellar wind. Therefore, any search for the presence of magnetic fields should involve multi-epoch spectropolarimetric observations. Given the small number of observations for each individual star in the study of Alecian et al. (2013), several magnetic Herbig Ae/Be stars were probably missed. This is illustrated in their study of HD 35929, which only on one occasion out of five showed the presence of a magnetic field. Thus, it is likely that a considerable number of the supposedly non-magnetic stars are also magnetic.

While previous magnetic surveys of Herbig Ae/Be stars have so far mainly targeted the detection of the magnetic fields, multi-epoch spectropolarimetric observations to understand their structure with respect to the stellar rotation axis and the disk orientation are important to monitor the targets over a significant number of nights to sample their rotation periods. However, until now, the magnetic field geometry was constrained only for two Herbig Ae/Be stars, the late Herbig Be star V380 Ori (Alecian et al. 2009) and the already discussed Herbig Ae star HD 101412 (Hubrig et al. 2011a). V380 Ori has a spectral type B9 and has been studied in great detail using multiwavelength observations (e.g., Hamann & Persson 1992; Rossi et al. 1999; Stelzer et al. 2006). It has a close infrared companion, with a separation of 0.15″ at PA 204° (Leinert et al. 1997). More recently, Reipurth et al. (2013) reported that V380 Ori is a hierarchical quadruple system with a fourth component at a distance of 8.8″ and position angle 120.4°. Using ESPaDOnS and Narval spectra, Alecian et al. (2009) detected the presence of a dipolar magnetic field with a polar strength of 2.12 ± 0.15 kG on the surface of the chemically peculiar primary. The authors also reported that the ESPaDOnS and Narval spectra of V380 Ori are heavily contaminated with strong, broad and variable emissions, but no periodicity of the observed variations could be established. The variability of the Stokes V signatures in the spectra of the chemically peculiar primary indicated a rotation period of about 4.3 days. Alecian et al. (2009) found that the primary in the V380 Ori system is itself a spectroscopic binary with a period of 104 days, with the secondary being a massive T Tauri star. Since no periodicity was found in the behaviour of the emission in hydrogen, helium, calcium and oxygen lines (the lines determining the Herbig Ae/Be nature), it is quite possible that the chemically peculiar component with the detected dipolar magnetic field is not a Herbig star, but already a star at an advanced age, probably on the zero-age main sequence (ZAMS), and that the Herbig Be status of the primary is merely based on the appearance of emission in the above mentioned lines belonging to the T Tauri component.

A similar scenario was found to be valid for the double-star system HD 72106, which showed the presence of emission in the Hα Balmer line of the combined spectrum of the system. In this system, located in the constellation Vela, the Herbig Ae/Be companion is separated from the primary by 0.805″ (Perryman et al. 1997). Vieira et al. (2003) suggested that the observed weak Hα emission and a small contribution from dust in the spectral energy distribution of HD 72106B was due to HD 72106B being an evolved Herbig Ae/Be star that has cleared most of its circumstellar envelope. Wade et al. (2005) announced the detection of a weak longitudinal magnetic field in the primary, assuming that the primary HD 72106A is a typical Herbig Ae/Be star. However, the study of Folsom et al. (2008) revealed that HD 72106A is a typical chemically peculiar young Bp star possessing a globally ordered magnetic field.

Since about 70% of the Herbig Ae/Be stars appear in binary/multiple systems (e.g., Baines et al. 2006), special care has to be taken in assigning the measured magnetic field to the particular component in the Herbig Ae/Be systems. A similar uncertainty in the evolutionary state also applies to the magnetic field detection in the system HD 200775 reported by Alecian et al. (2008). The authors announced the discovery of a dipolar magnetic field in the hot B3 Herbig Be star HD 200775, which is the primary in the double-lined spectroscopic binary system. However, the magnetic field was discovered not in the component possessing a circumstellar disk and dominating the Hα emission, so that the evolutionary status of the B3 primary component is unclear yet (Benisty et al. 2013).

6.3 Magnetospheric Accretion in Herbig Ae/Be Stars

Herbig Ae/Be stars have properties similar to classical T Tauri stars, for instance having emission lines, UV-excess and a lower surface gravity than main sequence stars (Hamann & Persson 1992) and they are usually identified by an infrared excess from circumstellar disks (e.g., Meeus et al. 2001). Because their envelopes are radiative, no magnetic field was expected to be generated in Herbig Ae/Be stars and thus no magnetically controlled accretion.

In the magnetospheric accretion scenario, stellar magnetic fields truncate the inner disk at a few stellar radii (e.g., Romanova et al. 2012) and cause gas flows from the truncated radius onto the star via magnetic field lines forming an accretion shock at the impact region (Koenigl 1991). Thus, the primary requirement for magnetospheric accretion to apply is the presence of a strong magnetic field capable of truncating the inner disk and channelling material toward the stellar surface. Bessolaz et al. (2008) addressed the issue of the disk truncation radius and its localisation as a function of the disk accretion rate and stellar dipole field parameters in T Tauri stars and concluded that the truncation radius decreases when the magnetic strength decreases and the mass accretion rate increases. The weakness of the observed magnetic fields in Herbig Ae/Be stars put into question the current understanding of the magnetospheric accretion process in these stars.

Accretion rates of 38 Herbig Ae/Be stars were determined from measurements of the UV-excess and Hα line width by Mendigutía et al. (2011), who reported that the Balmer excesses and Hα line widths of Herbig Ae stars can be interpreted within the context of magnetospheric accretion. Ababakr et al. (2017) presented Hα linear spectropolarimetry of a sample of 56 Herbig Ae/Be stars. A change in linear polarisation across this line was detected in 42 (75%) objects, indicating that the circumstellar environment around these stars on small spatial scales has an asymmetric structure, which is typically identified with a disk. A second outcome of their research was the confirmation that Herbig Ae stars are similar to T Tauri stars in displaying a line polarisation effect, while depolarisation is more common among Herbig Be stars. The mass accretion rate increases as a function of stellar mass (e.g., Fairlamb et al. 2015, 2017), and a different behaviour for objects with masses below and above the mass range 2.5–4.0 $M_\odot$ was detected by Wichittanakom et al. (2020). This break is consistent with the major finding in Fairlamb et al. (2015) that the relationship between mass accretion rate and stellar mass shows a break around the boundary between Herbig Ae and Herbig Be stars. Similar to the results of the work of Fairlamb et al. (2015), Wichittanakom et al. (2020) found indications that the mass accretion does decline with age as expected from basic theoretical considerations (e.g., Gorti & Hollenbach 2009), i.e., in terms of evolutionary age, the younger objects have higher accretion rates than older objects.

The dependence of the mass accretion rate on the age was already presented in the study of a sample of 25 Herbig Ae/Be stars by Hubrig et al. (2009), who reported that the highest mass accretion rates tend to be found in younger stars. The values of $\dot{M}_{\rm acc}$ in this study were taken from the work of Garcia Lopez et al. (2006), who derived $\dot{M}_{\rm acc}$ from the measured luminosity of the Brγ emission line using the correlation between L(Brγ) and the accretion luminosity $L_{\rm acc}$, established by Muzerolle et al. (1998a) and Calvet et al. (2004). The correlation used by Garcia Lopez et al. (2006) was empirical, without any assumption on the origin of the Brγ line. The dependence of the mass accretion rate on age for the sample of Herbig Ae/Be stars studied by Hubrig et al. (2009) is presented in Figure 6.12.

Magnetospheric accretion models describe the interaction between a dipolar stellar magnetic field and a surrounding accretion disk assuming pressure equilibrium. This analytical approach yields equations that relate the magnetic field strength to the system parameters such as mass, radius, accretion rate and rotation period (e.g., Koenigl 1991; Shu et al. 1994). Hubrig et al. (2009) examined the

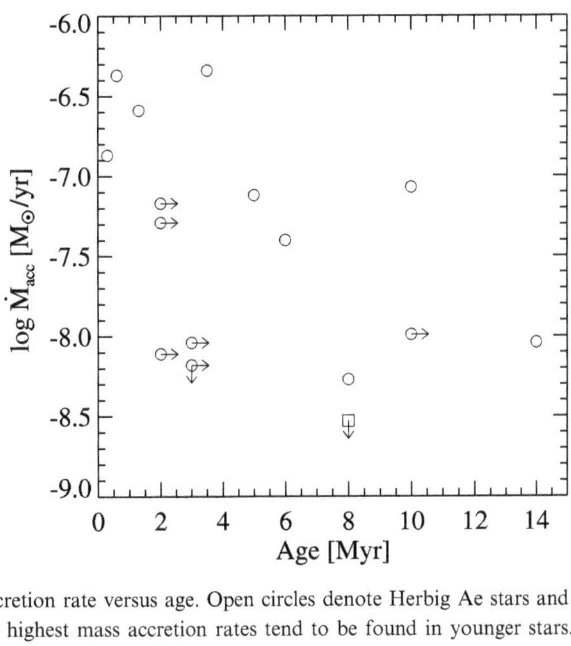

Figure 6.12. Mass accretion rate versus age. Open circles denote Herbig Ae stars and open squares indicate debris disk stars. The highest mass accretion rates tend to be found in younger stars. Credit: Hubrig et al. (2009), reproduced with permission © ESO.

magnetospheric accretion scenario for their sample of Herbig Ae stars using the expression for the surface equilibrium field given by Koenigl (1991). This relation can be presented in the form adopted by Johns-Krull et al. (1999):

$$B_{\rm eq}\,[{\rm kG}] = 3.43 \cdot \left(\frac{\epsilon}{0.35}\right)^{7/6} \cdot \left(\frac{\beta}{0.5}\right)^{-7/4} \cdot \left(\frac{M_*}{M_\odot}\right)^{5/6} \cdot \left(\frac{\dot{M}}{10^{-7} M_\odot/{\rm yr}}\right)^{1/2} \cdot \left(\frac{R_*}{R_\odot}\right)^{-3} \cdot \left(\frac{P_*}{\rm d}\right)^{7/6}, \tag{6.1}$$

where $B_{\rm eq}$ denotes the strength of the equatorial magnetic field in kG, the parameter ϵ expresses the stellar rotation rate in terms of rotation at the inner boundary of a truncated accretion disk, the parameter β describes how effectively the stellar magnetic field couples to the inner regions of the disk, M_* denotes the stellar mass in solar masses, $\dot{M}$ the accretion rate in 10^{-7} solar masses per year, R_* the stellar radius in solar radii and P_* the stellar rotation period in days. Johns-Krull (2007) has shown that for the case of cTTS, the models developed by different investigators make consistent predictions on the magnetic field strength.

The correlation between mass accretion rate and measured longitudinal magnetic field for the sample of Herbig Ae/Be stars from Hubrig et al. (2009) is presented in Figure 6.13. In this figure, the dotted line represents the prediction from Equation (6.1) for a canonical star with $M_* = 2.5\,M_\odot$, $R_* = 2.5\,R_\odot$ and $P_* = 0.5$ day.

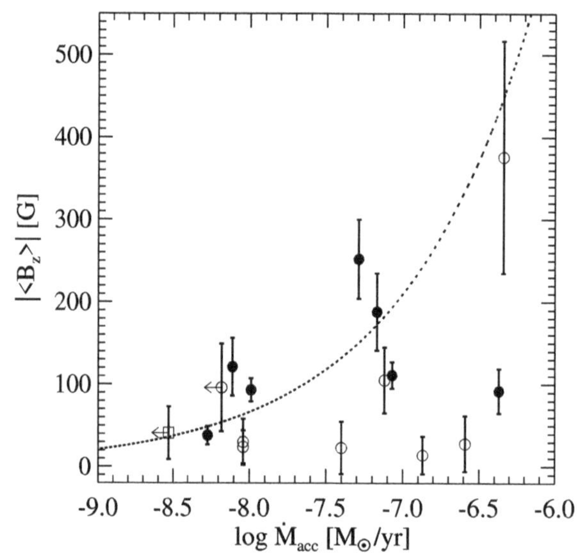

Figure 6.13. The strength of the longitudinal magnetic field as a function of accretion rate. Overplotted is the relation predicted by magnetospheric accretion models (see Equation (6.1)) for assumed stellar parameters of $M_* = 2.5\,M_\odot$, $R_* = 2.5\,R_\odot$ and $P_* = 0.5$ day. Filled circles denote Herbig Ae stars with a 3σ magnetic field detection, while open circles denote Herbig Ae stars with non-detections. Squares denote stars with debris disks, of which none has a 3σ magnetic field detection. Credit: Hubrig et al. (2009), reproduced with permission © ESO.

The presented trend between the strength of the longitudinal magnetic field and the accretion rate qualitatively supports the magnetospheric accretion model for Herbig Ae stars. It should be noted that Equation (6.1) refers to the equatorial dipole field, while the measured values represent an average of the longitudinal component over the stellar surface. These values also depend on the viewing angle.

Interferometric searches for magnetospheres in the near-infrared did not come to a final conclusion. Kraus et al. (2008) observed five Herbig Ae/Be stars with AMBER on the Very Large Telescope Interferometer and found only in HD 98922—then not known to be magnetic—a Brγ line-emitting region compact enough to be compatible with a magnetosphere. A more recent work of Järvinen et al. (2019a) using HARPSpol observations of HD 98922 confirmed the presence of a weak magnetic field in this star (see Figure 6.11). For the other four sources in the study, among them the magnetic SB2 system HD 104237, Kraus et al. (2008) found larger sizes consistent with an extended stellar wind or a disk-wind. It is not clear why the interferometric observations by Kraus et al. (2008) failed to detect the magnetosphere in the system HD 104237. In contrast, as reported above, studying magnetospheric accretion in HD 58647, Kurosawa et al. (2016) modelled continuum and Brγ AMBER data and found that a disk-wind plus a small magnetosphere explain all measurables. Indeed, the presence of a magnetic field in this star was recently reported by Järvinen et al. (2019a).

Several Hα spectro-interferometric studies of Herbig Ae/Be stars were carried out with the CHARA array. Perraut et al. (2016) found evidence for a disk-wind coming from 0.3 au and for a magnetosphere in AB Aur. Benisty et al. (2013) saw a disk-wind on a scale of 0.2–0.6 au in MWC 361. Mendigutía et al. (2017) determined the sizes of the Hα emitting regions smaller than 0.21 au and 0.12 au for HD 179218 and HD 141529, respectively, which are likely due to magnetospheric accretion. The Hα profile of HD 141569 could not be fitted due to the high rotational velocity of this object. Putting the CHARA sample together, Mendigutía et al. (2017) concluded that a wider variety of scenarios is needed to explain the Hα emission in Herbig Ae/Be stars, including contributions coming from compact or extended sources, from a disk, from the accretion flows and from winds.

Spectropolarimetric observations of photospheric lines are usually used to determine the geometry of the global magnetic field, while spectropolarimetry of accretion diagnostic lines, such as He I λ5876, He I λ10830, the Na I doublet and the hydrogen lines, probes the accretion topology in the accretion columns, which are perturbed by interactions with the disk. Reiter et al. (2018) studied single-epoch He I λ10830 line profiles for 64 Herbig Ae/Be stars with magnetic field measurements mentioned in the work of Alecian et al. (2013), including five Herbig Ae/Be stars identified as magnetic in order to test the presence of magnetospheric accretion. The authors examined the shapes of He I λ10830 profiles for appearance of redshifted or blueshifted absorptions, assuming that blueshifted absorptions are produced by colder gas moving toward the observer in a wind or outflow and the redshifted absorptions occur when colder material moves toward the star, and thus away from the observer. The uniqueness of this diagnostic line derives from the metastability of this transition and makes it a good indicator of wind and funnel flow geometry

(Edwards et al. 2006). Further, according to Edwards et al. (2006), this He I line appears in emission for stronger mass accretion rates and in net absorption for low mass accretion rates. Reiter et al. (2018) reported that of the five Herbig Ae/Be stars with a magnetic field detection, one showed redshifted absorption, indicative of infall, and two showed blueshifted absorption, tracing mass outflow. The fraction of redshifted and blueshifted absorption profiles in the non-magnetic Herbig Ae/Be stars was however remarkably similar, suggesting that a stellar magnetic field does not affect gas kinematics traced by the He I λ10830 line.

It is not surprising that the authors did not find any correlation of the He I λ10830 line profile morphology with the presence of a magnetic field. Since Reiter et al. (2018) used single-epoch observations, this study completely ignored the presence of spectral and magnetic variability usually observed in Herbig Ae/Be stars. Unless the observer directly looks onto the rotation pole, or the obliquity angle between the rotation axis and the magnetic field axis is zero, rotational modulation of magnetic field strengths and the line profile shapes is expected. As an example, a strong variability of the He I λ10830 line profiles over the rotation period of the strongly magnetic Herbig Ae star HD 101412 is presented in Figure 6.14. High-resolution

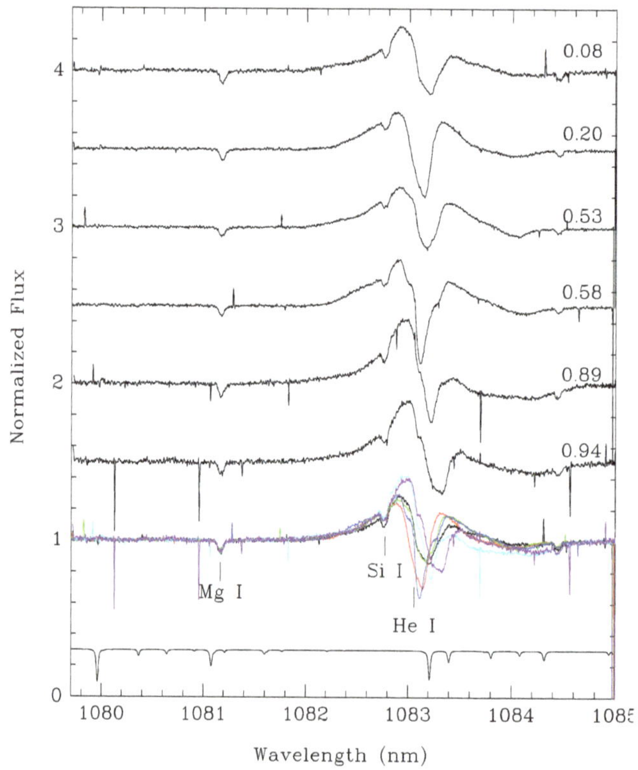

Figure 6.14. The variability of the He I λ10830 line profile over the rotation period of 42 days in the spectra of HD 101412. The telluric spectrum presented at the bottom of the figure was removed from the observed spectra. The second spectrum from the bottom presents all six spectra overplotted. Credit: Hubrig et al. (2012), reproduced with permission © ESO.

CRIRES spectra of this star were collected on six different rotation phases (Hubrig et al. 2012).

It was also not clear to which aspect of the (unknown) magnetic field geometry the He I λ10830 line observed by Reiter et al. (2018) on single epochs belonged. Obviously, understanding the interaction of the magnetic field with the stellar environment is possible only in observations with good coverage of the rotation cycle. Furthermore, a sample of five magnetic stars is not sufficient to conclusively demonstrate that the He I λ10830 profile shapes in magnetic objects are, or are not, different compared to those observed in non-magnetic objects.

The excellent potential of diagnostic He I and hydrogen lines to study magnetospheric accretion in Herbig Ae/Be stars was demonstrated in the work of Schöller et al. (2016). The authors used near-infrared spectroscopic observations of HD 101412 to test the magnetospheric character of its accretion. As already discussed in Section 6.2, this star possesses the strongest magnetic field measured in any Herbig Ae star so far, with a surface magnetic field $\langle B \rangle$ up to 3.5 kG (Hubrig et al. 2011a). The phase diagram for the measurements using FORS 2 spectropolarimetric observations is presented in Figure 6.6. This star has well-determined stellar parameters: the rotation period $P_{\rm rot} = 42.076 \pm 0.017$ day, the inclination angle of the rotation axis to the line of sight $i = 80 \pm 7°$ and the obliquity angle, i.e., the angle between the rotation axis and the axis of the magnetic dipole, $\beta = 84 \pm 13°$. As the He I λ10830 and Paγ lines are usually considered to form in the accretion regions of PMS stars, Schöller et al. (2016) analysed their behaviour using 30 near-IR spectra acquired with the VLT instruments CRIRES and X-shooter. Spectral parameters of these diagnostic lines used for a quantitative analysis and the resulting periodograms are presented in Figure 6.15. Schöller et al. (2016) showed that the temporal behaviour of the He I λ10830 and Paγ lines can be explained by rotational modulation of accreting gas with a period $P = 20.53 \pm 1.68$ days, i.e., the discovered modulation of these diagnostic lines is governed by the rotation period of HD 101412 and the observed spectroscopic variability can be explained within the magnetic geometry established earlier from magnetic field measurements. The discovery of this period, about half of the magnetic rotation period $P_{\rm rot} = 42.076$ days, indicates that the accreted matter falls onto the star in regions close to the magnetic poles intersecting the line of sight twice during the rotation cycle. An artist's impression of the magnetospheric accretion in HD 101412 is presented in Figure 6.16.

6.4 Magnetic Field Studies of Vega-like Stars

Since the discovery of the large infrared excess at wavelengths longwards of 12 μm from Vega (Aumann et al. 1984), more than one hundred main sequence stars have been found to have a similar infrared excess in the IRAS wavebands and have been called Vega-like stars. They are predominantly spectral type A main sequence stars with continuum emission at far-infrared wavelengths in excess of that expected from the photospheres of stars of their spectral type. The excess flux is ascribed to thermal emission from orbiting dust grains with a temperature of about 100 K (Aumann 1985).

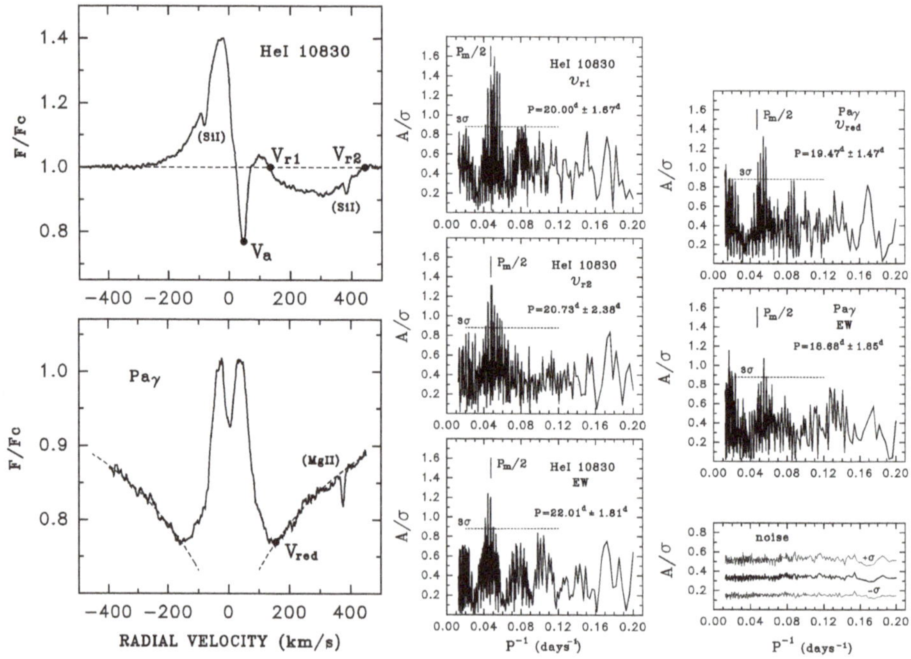

Figure 6.15. *Left panel:* spectral parameters of the He I 10830 and Paγ line profiles used in the quantitative analysis of HD 101412. *Right panel:* A/σ periodograms for different line parameters. Significance levels of 3σ are indicated by the dashed lines. Short vertical lines indicate the value corresponding to half of the magnetic rotation period ($P_{\rm rot}/2 = 21.038$ days). In each plot, the detected periods and their uncertainties are indicated. Credit: Schöller et al. (2016), reproduced with permission © ESO.

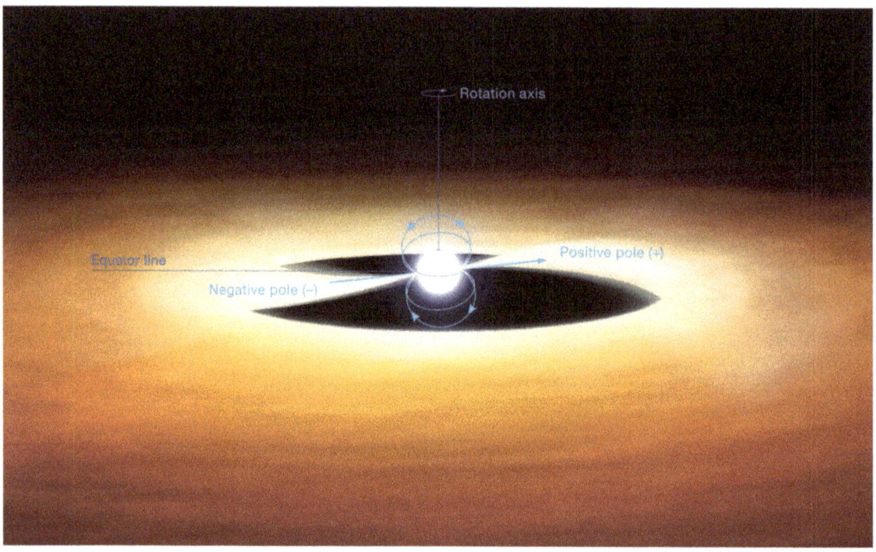

Figure 6.16. Artist's impression of the magnetospheric accretion in HD 101412, looking at the magnetic equator. Credit: Schöller et al. (2016), reproduced with permission © ESO.

In the evolutionary context, Herbig Ae/Be stars can be considered as the precursors of Vega-like stars. The dereddened optical colours of Vega-like stars are consistent with photospheric emission from main sequence stars. However, for a number of Vega-like stars, the mid- and far-IR photometry revealed excess IR emission with colours resembling those of Herbig Ae/Be systems (e.g., Sheret et al. 2004). One of the prototypes of Vega-like stars, β Pic, shows evidence of a complex stellar environment with dense and hot regions. It was previously argued that the observed broad emission from the highly ionised species C III and O IV may originate from a solar-like extended chromosphere or from magnetospheric accretion (Deleuil et al. 2001). If solar-like activity is responsible for the observed highly ionised species, a theoretical model to explain the origin and evolution of these phenomena remains to be built.

A first attempt to measure a magnetic field in the Vega-like star β Leo was carried out about 40 years ago by Landstreet (1982), who used a two-channel photoelectric cell polarimeter as a Balmer-line Zeeman analyser. The author failed to detect the presence of a longitudinal magnetic field in this star measuring $\langle B_z \rangle = -80 \pm 65$ G. No magnetic field detection at a significance level of 3σ could be diagnosed in any of seven Vega-like stars studied by Hubrig et al. (2006) using low-resolution FORS 1 spectropolarimetric observations. The most accurate determination of a magnetic field, at a 2.6σ significance level, was performed for the Vega-like star ι Cen, for which the authors measured $\langle B_z \rangle = -77 \pm 30$ G. For the star β Pic, which was reported to show conspicuous signs of chromospheric activity (Bouret et al. 2002), the longitudinal magnetic field was measured only at a significance level of $\sim 1.5\sigma$ with $\langle B_z \rangle = -79 \pm 53$ G. A search for magnetic fields using low-resolution FORS 1 spectropolarimetric observations was carried out for three more Vega-like stars, HD 172555, HD 215789, and HD 224392 by Schröder et al. (2008). Also for these stars magnetic field measurements yielded no detections. The highest significance level of 2.6σ with $\langle B_z \rangle = 134 \pm 51$ G was achieved for HD 224392 using the entire spectra.

The presented non-detections in Vega-like stars indicate that longitudinal magnetic fields in these stars should be rather weak with field strengths below 100 G. The detection limit also depends on the structure of their magnetic fields, which is unknown.

6.5 Magnetic Field Origin

The origin of stellar magnetic fields in Herbig Ae/Be stars is still not properly understood. Magnetic fields in these stars might be fossils of the early star formation epoch, in which the magnetic field of the parental magnetised core was compressed into the innermost regions of the accretion disks (e.g., Banerjee & Pudritz 2006). Alternatively, Tout & Pringle (1995) proposed a non-solar dynamo that could operate in rapidly rotating A-type stars based on rotational shear energy.

Hubrig et al. (2009) studied a sample of 25 Herbig Ae/Be stars and six debris disk stars with the aim to determine whether magnetic field properties of these stars are correlated with other observed properties such as mass accretion rate, accretion disk inclination, companions, silicates, PAHs (polycyclic aromatic hydrocarbons), or show a correlation with age and X-ray emission as expected from the decay of a remnant dynamo. Since a different accretion scenario was suggested for Herbig Be stars with respect to Herbig Ae stars, the sample was restricted to Herbig Ae and late Herbig Be stars. As already discussed in Section 6.3, no simple dependence of the magnetic field strength on the mass accretion rate was found, but the range of the observed field values qualitatively supported the expectations from magnetospheric accretion models with dipole-like field geometries. The analysis of X-ray luminosities collected from the literature showed a hint for an increase of the magnetic field strength with the level of the X-ray emission, also supported by the correlation analysis, which yielded >90% probability that a correlation is present between $\langle B_z \rangle$ and $\log L_X$. This can suggest a dynamo mechanism responsible for the coronal activity in Herbig Ae stars. The strength of the magnetic field plotted versus $\log L_X$ is presented in Figure 6.17.

The authors found clear indications for a trend toward stronger magnetic fields in younger Herbig Ae stars, confirmed by statistical tests. The plot displaying the strength of the longitudinal magnetic field as a function of age is shown in Figure 6.18. Although it has been frequently mentioned in the literature that magnetic Herbig Ae stars are potential progenitors of the magnetic Ap stars (e.g., Stępień & Landstreet 2002; Catala et al. 2007; Wade et al. 2005), it is obvious from Figure 6.18 that stronger magnetic fields appear in very young Herbig Ae stars, and that the magnetic fields become very weak at the end of their PMS life. These results are in accordance with the conclusions of Hubrig et al. (2000, 2005, 2007a) that magnetic fields in Ap stars with masses lower than 3 $M_\odot$ are rarely found close to the ZAMS and that kG

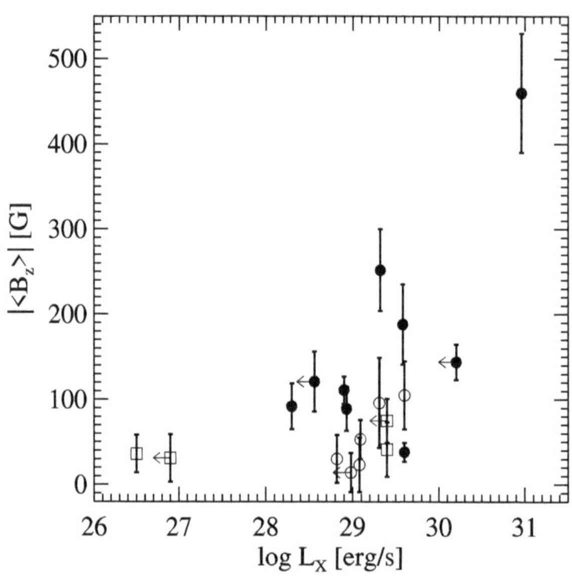

Figure 6.17. The strength of the longitudinal magnetic field plotted over the X-ray luminosity. Filled circles denote Herbig Ae stars with a 3σ magnetic field detection, while open circles denote Herbig Ae stars with non-detections. Squares denote stars with debris disks, of which none has a 3σ magnetic field detection. Credit: Hubrig et al. (2009), reproduced with permission © ESO.

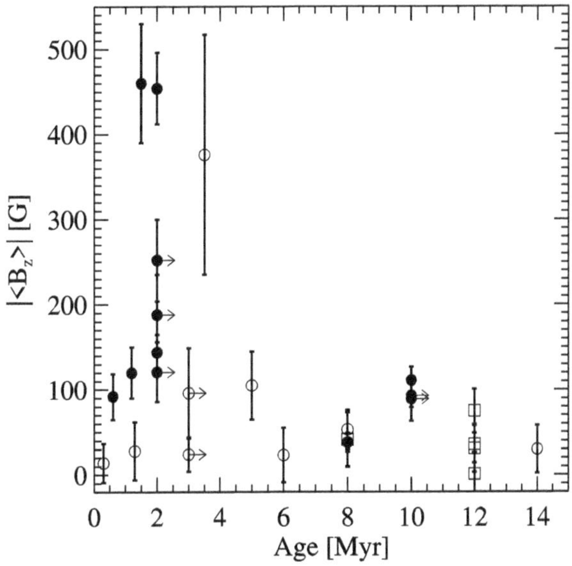

Figure 6.18. The strength of the longitudinal magnetic field as a function of age. The symbols are identical to Figure 6.17. Credit: Hubrig et al. (2009), reproduced with permission © ESO.

magnetic fields appear in A stars already evolved from the ZAMS. In contrast, magnetic Bp stars with masses $M < 3\ M_{\odot}$ seem to be concentrated closer to the ZAMS.

Both the magnetic field strength and the X-ray emission showed hints of a decline with age in the range of ~2 – 14 Myr probed by the sample, supporting a dynamo mechanism that decays with age. Importantly, the studied stars seemed to obey the universal power-law relation between magnetic flux and X-ray luminosity established for the Sun and main-sequence active dwarf stars (Pevtsov et al. 2003). It is of interest that Alecian (2014) suggested that the magnetic properties of A and B-type stars were shaped before the Herbig Ae/Be evolutionary phase. Using PMS evolutionary tracks calculated with the CESAM code (Morel 1997), the authors concluded that even stars above 3 $M_{\odot}$ undergo a purely convective phase before reaching the birthline. It is therefore quite possible that the weak magnetic fields detected in Herbig Ae/Be stars (Hubrig et al. 2015a) are just leftovers of the magnetic fields generated by dynamos during these convective phases

The study of Hubrig et al. (2009) did not indicate any correlation of the strength of the longitudinal magnetic field with disk orientation or disk geometry, but there was a hint that magnetic fields are most frequent in stars with flared disks and hot, inner gas. Also the membership in binary or multiple systems did not seem to have any impact on the presence of magnetic fields. Interestingly, no trend with rotation velocity or rotation period was found. Although the obtained results provided new clues, the observations presented in the work of Hubrig et al. (2009) were still inconclusive as to the magnetic field origin. Future studies of stellar properties of magnetic Herbig stars should involve a larger and more representative sample to determine the existing relations at a higher confidence level.

Noteworthy, recent observations of magnetic Ap/Bp stars support the magnetic field origin scenario that requires mergers to complete early at the PMS stage. According to Ferrario et al. (2009), at least one of the merging stars must be on the Henyey part of the PMS track toward the end of its contraction to the main sequence. The merger outcome then becomes observable as a magnetic Herbig Ae/Be star, implying that there should be almost no magnetic star in close Herbig Ae/Be and Ap/Bp binaries. On the other hand, according to Tutukov & Fedorova (2010), it is possible that magnetic Herbig Ae/Be stars and Ap/Bp stars are formed through a completely different channel. A system formed through a synchronised binary channel could undergo an external perturbation leading to its components moving further apart, with an increased orbital period, and building a wide binary system. In agreement with these scenarios, the Herbig Ae/Be binarity studies by Wheelwright et al. (2010) and Duchêne (2015) and the binarity study of Ap/Bp stars by Mathys (2017) clearly indicate the lack of close binary systems among these types of objects.

The analysis of HARPSpol observations of the rare close SB2 system HD 104237 with $P_{\rm orb}$ = 19.859 days confirmed that both components, the Herbig Ae star and the lower mass T Tauri star, possess magnetic fields (Järvinen et al. 2019b). The results of the application of the LSD technique to all 88 HARPSpol spectra of this system are presented in Figure 6.19. As illustrated in this figure, the LSD spectra appear very noisy. Therefore, to more accurately characterise the magnetic field variability in both components, the dedicated Singular Value Decomposition technique (Carroll et al. 2012) was applied to the observations obtained at phases where both components were well separated. The longitudinal magnetic field of the primary, the Herbig Ae star is weak and is only slightly changing from 47 G to 72 G. In the phases, where the components are blended, the Stokes V profiles were assumed to be dominated by the magnetic field of the T Tauri secondary. The rather strong longitudinal magnetic field of the T Tauri star was estimated on multiple epochs and showed strong variability. In orbital phases where the T Tauri component in the composite spectrum was well separated from the Herbig Ae star, $\langle B_z \rangle$ showed a field strength decrease from 609 G to 124 G within 7.6 h. The variability of the equivalent widths of the primary component indicated a rotation period of 4.337 17 ± 0.003 16 day (104 ± 0.08 hours), which is in good agreement with the value (100 ± 5 h) reported by Böhm et al. (2006), who estimated the period from the analysis of the Hα line. No rotation period could be determined for the secondary due to its low flux contribution in the composite spectrum.

Recent observations of the two Herbig Ae/Be SB2 binaries AK Sco and HD 95881 using HARPSpol showed that magnetic fields are present also in binary Herbig Ae/Be systems. Employing the multi-line singular value decomposition (SVD) method, Järvinen et al. (2018) detected a mean longitudinal magnetic field $\langle B_z \rangle = -83 \pm 31$ G in the secondary component of AK Sco. For HD 95881, the authors measured at two different observing epochs $\langle B_z \rangle = -93 \pm 25$ G and $\langle B_z \rangle = 105 \pm 29$ G. For all detections the FAP was smaller than 10^{-5}, indicating definite detections (see Section 2.10).

In Figure 6.20, the calculated Stokes I and Stokes V using the SVD method, and the diagnostic null N profile for both systems, AK Sco and HD 95881, are presented.

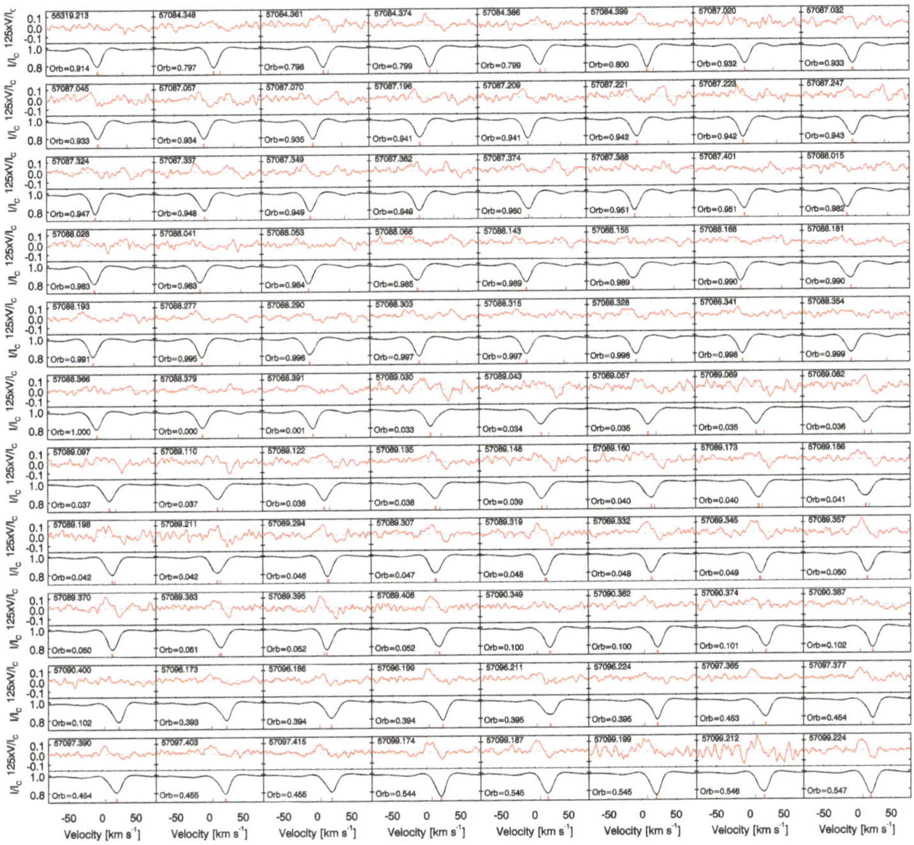

Figure 6.19. Stokes *I* (*bottom panels*) and Stokes *V* (*top panels*) Least-Square-Deconvolution profiles for 88 observations of the SB2 system HD 104237 consisting of a Herbig Ae primary and a T Tauri secondary. The Stokes *V* profiles have been amplified by a factor of 125. The primary is marked with thick red ticks and the secondary with thin blue ticks below the Stokes *I* profiles. Credit: Figure 2 from Järvinen et al. (2019b).

The diagnostic *N* profiles are usually used to identify spurious polarisation signatures. Since, according to Alencar et al. (2003), both components in AK Sco are expected to be tidally synchronised, the detected Zeeman feature observed at the epoch 2017 June 6 corresponds to the region of the stellar surface of the secondary component facing permanently the primary component, meaning that the magnetic field geometry in this component is tidally locked and is closely related to the position of the primary component. Similar tidally locked magnetic field geometries were detected in the systems HD 98088 and HD 161701 having a magnetic Ap star as a component (Abt et al. 1968; Hubrig et al. 2014a).

It is expected that studies of the presence of a magnetic field and of the alignment of the magnetic axis to the orbital radius vector in Herbig stars that are members of binary systems can have the potential to disclose the mechanism of the field generation in Herbig Ae/Be stars. Future studies are necessary to probe the correlations between the binary properties, magnetic field structure and strength, and abundance inhomogeneities.

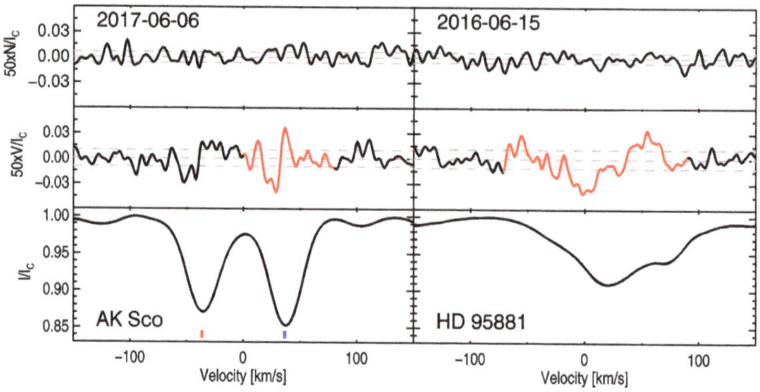

Figure 6.20. Examples of Stokes I, (*bottom*), Stokes V (*middle*), and diagnostic null (N) (*top*) profiles obtained for the Herbig SB2 systems AK Sco and HD 95881. The detected Zeeman features are highlighted in red. The horizontal dashed lines indicate the $\pm 1\sigma$-ranges. For AK Sco, the primary component is marked with a red tick and the secondary with a blue tick.

6.6 Future Prospects in Magnetic Studies of Herbig Ae/Be Stars

It appears that the as yet small number of magnetic Herbig Ae/Be stars is due to the weakness of the magnetic fields and/or the large measurement errors. While magnetic surveys of Herbig Ae/Be stars so far have mainly targeted the detection of magnetic fields, knowledge of the magnetic field structure combined with an accurate determination of the stellar fundamental parameters are indispensable to constrain theories on star formation and magnetospheric accretion in intermediate-mass stars. Further, it is necessary to understand the three-dimensional structure of magnetic fields with respect to stellar rotation axis and disk orientation. The Doppler-shifted spectropolarimetric contributions from photospheric and circumstellar environmental diagnostic lines will allow to apply the technique of Zeeman–Doppler tomography to determine the correspondence between the magnetic field structure and the radial density and temperature profiles. In a circumstellar environment, such as a wind or an accretion disk, lines may form over a relatively large volume, and the field topology may be therefore complex not only in latitude and azimuth, but in radius as well. The motion of the circumstellar gas is expected to be governed by the magnetic field, especially in the regions of the stellar wind, which is generally assumed to be accelerated by a magnetic centrifuge. The study of high-resolution polarimetric spectra of HD 101412 obtained with HARPSpol by Järvinen et al. (2015b) revealed that different line lists used in the SVD method for Stokes profile reconstructions yielded differences in both the shape of the Stokes V profiles and the field strengths (see Section 3.3). Distinct features appeared as a second maximum in the red wings of the SVD Stokes V profiles obtained at a few epochs. Furthermore, these profiles seemed to be slightly shifted to the blue by about 4 km s^{-1}. The observed behaviour of Zeeman features was interpreted in the context of the impact of circumstellar matter and element abundance inhomogeneities on the measurements of the magnetic field. The ultimate goal of future studies of Herbig Ae/Be stars is to understand the accretion geometry and the disk–magnetosphere interaction based on Zeeman–Doppler tomographic studies.

Studies of the field structure, its stability across the PMS tracks, the impact of magnetic fields on the evolution of the rotation rates and possible correlations between evolutionary state and other stellar properties are furthermore of fundamental importance for magnetohydrodynamical simulations of the star formation process. Studying the magnetic field structure and its evolution will lead to a deep insight into the complex interaction between the stellar magnetic field, the accretion disk and the stellar wind: it will provide crucial additional information to test the predictions of existing theories on the origin of the magnetic field. Moreover, since current multiwavelength observations reveal that the protoplanetary disks of Herbig Ae/Be stars frequently show intriguing structures both in scattered light and thermal IR emission, which are assumed to be related to planet formation processes and planet–disk interactions, future studies aimed at the characterisation of the magnetic field structure in Herbig Ae/Be stars will clarify the role of magnetic fields in the planet formation processes.

References

Ababakr, K. M., Oudmaijer, R. D., & Vink, J. S. 2017, MNRAS, 472, 854

Abt, H. A., Conti, P. S., Deutsch, A. J., & Wallerstein, G. 1968, ApJ, 153, 177

Alecian, E., Catala, C., Wade, G. A., et al. 2008, MNRAS, 385, 391

Alecian, E., Wade, G. A., Catala, C., et al. 2009, MNRAS, 400, 354

Alecian, E., Wade, G. A., Catala, C., et al. 2013, MNRAS, 429, 1001

Alecian, E. 2014, in Putting A Stars into Context: Evolution, Environment, and Related Stars, ed. O. Mathys, R. Monier, & G. M. Wahlgren (Moscow: Pero Publishing House), 84

Alencar, S. H. P., Melo, C. H. F., Dullemond, C. P., et al. 2003, A&A, 409, 1037

Aumann, H. H., Gillett, F. C., Beichman, C. A., et al. 1984, ApJL, 278, L23

Aumann, H. H. 1985, PASP, 97, 885

Böhm, T., Catala, C., Balona, L., & Carter, B. 2004, A&A, 427, 907

Böhm, T., Dupret, M. A., & Aynedjian, H. 2006, MmSAI, 77, 362

Baines, D., Oudmaijer, R. D., Porter, J. M., & Pozzo, M. 2006, MNRAS, 367, 737

Banerjee, R., & Pudritz, R. E. 2006, ApJ, 641, 949

Benisty, M., Perraut, K., Mourard, D., et al. 2013, A&A, 555, A113

Bertrang, G. H. M., Flock, M., Keppler, M., et al. 2020, arXiv:2007.11565

Bessolaz, N., Zanni, C., Ferreira, J., Keppens, R., & Bouvier, J. 2008, A&A, 478, 155

Boccaletti, A., Di Folco, E., Pantin, E., et al. 2020, A&A, 637, L5

Bouret, J. C., Deleuil, M., Lanz, T., et al. 2002, A&A, 390, 1049

Calvet, N., Muzerolle, J., Briceño, C., et al. 2004, AJ, 128, 1294

Carroll, T. A., Strassmeier, K. G., Rice, J. B., & Künstler, A. 2012, A&A, 548, A95

Catala, C., Bohm, T., Donati, J. F., & Semel, M. 1993, A&A, 278, 187

Catala, C., Alecian, E., Donati, J. F., et al. 2007, A&A, 462, 293

Cauley, P. W., & Johns-Krull, C. M. 2014, ApJ, 797, 112

Cowley, C. R., Hubrig, S., González, J. F., & Savanov, I. 2010, A&A, 523, A65

Deleuil, M., Bouret, J. C., Lecavelier des Etangs, A., et al. 2001, ApJL, 557, L67

Donati, J. F., Semel, M., Carter, B. D., Rees, D. E., & Collier Cameron, A. 1997, MNRAS, 291, 658

Duchêne, G. 2015, Ap&SS, 355, 291

Edwards, S., Fischer, W., Hillenbrand, L., & Kwan, J. 2006, ApJ, 646, 319
Fairlamb, J. R., Oudmaijer, R. D., Mendigutía, I., Ilee, J. D., & van den Ancker, M. E. 2015, MNRAS, 453, 976
Fairlamb, J. R., Oudmaijer, R. D., Mendigutia, I., Ilee, J. D., & van den Ancker, M. E. 2017, MNRAS, 464, 4721
Fedele, D., van den Ancker, M. E., Acke, B., et al. 2008, A&A, 491, 809
Ferrario, L., Pringle, J. E., Tout, C. A., & Wickramasinghe, D. T. 2009, MNRAS, 400, L71
Folsom, C. P., Wade, G. A., Kochukhov, O., et al. 2008, MNRAS, 391, 901
Garcia Lopez, R., Natta, A., Testi, L., & Habart, E. 2006, A&A, 459, 837
Garcia, P. J. V., Benisty, M., Dougados, C., et al. 2013, MNRAS, 430, 1839
Glagolevskij, Y. V., & Chountonov, G. A. 2001, in ASP Conf. Ser. 248, Magnetic Fields Across the Hertzsprung–Russell Diagram, ed. G. Mathys, S. K. Solanki, & D. T. Wickramasinghe (San Francisco, CA: ASP), 535
Gorti, U., & Hollenbach, D. 2009, ApJ, 690, 1539
Hamann, F., & Persson, S. E. 1992, ApJS, 82, 285
Hartmann, L., Hewett, R., & Calvet, N. 1994, ApJ, 426, 669
Herbig, G. H., & Bell, K. R. 1988, Lick Observatory Bulletin 1111, Third Catalog of Emission-Line Stars of the Orion Population 3 (Santa Cruz, CA: Lick Observatory)
Herbig, G. H. 1960, ApJS, 4, 337
Hubrig, S., North, P., & Mathys, G. 2000, ApJ, 539, 352
Hubrig, S., Schöller, M., & Yudin, R. V. 2004, A&A, 428, L1
Hubrig, S., Schöller, M., & North, P. 2005, in AIP Conf. Ser. 784, Magnetic Fields in the Universe: From Laboratory & Stars to Primordial Structures, ed. E. M. de Gouveia dal Pino, G. Lugones, & A. Lazarian (Melville, NY: AIP), 145
Hubrig, S., Yudin, R. V., Schöller, M., & Pogodin, M. A. 2006, A&A, 446, 1089
Hubrig, S., North, P., & Schöller, M. 2007a, AN, 328, 475
Hubrig, S., Pogodin, M. A., Yudin, R. V., Schöller, M., & Schnerr, R. S. 2007b, A&A, 463, 1039
Hubrig, S., Stelzer, B., Schöller, M., et al. 2009, A&A, 502, 283
Hubrig, S., Schöller, M., Savanov, I., et al. 2010, AN, 331, 361
Hubrig, S., Mikulášek, Z., González, J. F., et al. 2011a, A&A, 525, L4
Hubrig, S., Schöller, M., Ilyin, I., et al. 2011b, A&A, 536, A45
Hubrig, S., Castelli, F., González, J. F., et al. 2012, A&A, 542, A31
Hubrig, S., Ilyin, I., Schöller, M., & Lo Curto, G. 2013, AN, 334, 1093
Hubrig, S., Carroll, T. A., Gonzalez, J. F., et al. 2014, MNRAS, 440, L6
Hubrig, S., Carroll, T. A., Scholler, M., & Ilyin, I. 2015a, MNRAS, 449, L118
Järvinen, S. P., Carroll, T. A., Hubrig, S., et al. 2015b, A&A, 584, A15
Järvinen, S. P., Carroll, T. A., Hubrig, S., et al. 2018, ApJL, 858, L18
Järvinen, S. P., Carroll, T. A., Hubrig, S., Ilyin, I., & Schöller, M. 2019a, MNRAS, 489, 886
Järvinen, S. P., Carroll, T. A., Hubrig, S., et al. 2019b, MNRAS, 486, 5499
Johns-Krull, C. M., Valenti, J. A., & Koresko, C. 1999, ApJ, 516, 900
Johns-Krull, C. M. 2007, ApJ, 664, 975
Koenigl, A. 1991, ApJL, 370, L39
Kraus, S., Hofmann, K. H., Benisty, M., et al. 2008, A&A, 489, 1157
Krumholz, M. R., & Federrath, C. 2019, FrASS, 6, 7
Krumholz, M. R., & Matzner, C. D. 2009, ApJ, 703, 1352
Kurosawa, R., Kreplin, A., Weigelt, G., et al. 2016, MNRAS, 457, 2236

Landstreet, J. D. 1982, ApJ, 258, 639
Leinert, C., Richichi, A., & Haas, M. 1997, A&A, 318, 472
Mathys, G. 2017, A&A, 601, A14
Matt, S., & Pudritz, R. E. 2005, ApJL, 632, L135
McKee, C. F., & Ostriker, E. C. 2007, ARA&A, 45, 565
McKee, C. F., Stacy, A., & Li, P. S. 2020, MNRAS, 496, 5528
Meeus, G., Waters, L. B. F. M., Bouwman, J., et al. 2001, A&A, 365, 476
Mendigutía, I., Calvet, N., Montesinos, B., et al. 2011, A&A, 535, A99
Mendigutía, I., Oudmaijer, R. D., Mourard, D., & Muzerolle, J. 2017, MNRAS, 464, 1984
Morel, P. 1997, A&AS, 124, 597
Muro-Arena, G. A., Benisty, M., Ginski, C., et al. 2020, A&A, 635, A121
Muzerolle, J., Hartmann, L., & Calvet, N. 1998a, AJ, 116, 2965
Muzerolle, J., Hartmann, L., & Calvet, N. 1998b, AJ, 116, 455
Muzerolle, J., D'Alessio, P., Calvet, N., & Hartmann, L. 2004, ApJ, 617, 406
Najita, J. 2004, in ASP Conf. Ser. 323, Star Formation in the Interstellar Medium: In Honor of David Hollenbach, ed. D. Johnstone, F. C. Adams, D. N. C. Lin, D. A. Neufeeld, & E. C. Ostriker (San Francisco, CA: ASP), 271
Nealon, R., Cuello, N., Gonzalez, J.-F., et al. 2020, MNRAS, 499, 3857
Penz, T., Micela, G., & Lammer, H. 2008, A&A, 477, 309
Perraut, K., Dougados, C., Lima, G. H. R. A., et al. 2016, A&A, 596, A17
Perryman, M. A. C., Lindegren, L., Kovalevsky, J., et al. 1997, A&A, 500, 501
Pevtsov, A. A., Fisher, G. H., Acton, L. W., et al. 2003, ApJ, 598, 1387
Pineda, J. E., Szulágyi, J., Quanz, S. P., et al. 2019, ApJ, 871, 48
Pinte, C., van der Plas, G., Ménard, F., et al. 2019, NatAs, 3, 1109
Pinte, C., Price, D. J., Ménard, F., et al. 2020, ApJL, 890, L9
Reipurth, B., Bally, J., Aspin, C., et al. 2013, AJ, 146, 118
Reiter, M., Calvet, N., Thanathibodee, T., et al. 2018, ApJ, 852, 5
Romanova, M. M., Ustyugova, G. V., Koldoba, A. V., Wick, J. V., & Lovelace, R. V. E. 2003, ApJ, 595, 1009
Romanova, M. M., Ustyugova, G. V., Koldoba, A. V., & Lovelace, R. V. E. 2012, MNRAS, 421, 63
Rosner, R., Golub, L., & Vaiana, G. S. 1985, ARA&A, 23, 413
Rossi, C., Errico, L., Friedjung, M., et al. 1999, A&AS, 136, 95
Schöller, M., Pogodin, M. A., Cahuasquí, J. A., et al. 2016, A&A, 592, A50
Schröder, C., Hubrig, S., & Schmitt, J. H. M. M. 2008, A&A, 484, 479
Semel, M. 1989, A&A, 225, 456
Sheret, I., Dent, W. R. F., & Wyatt, M. C. 2004, MNRAS, 348, 1282
Shu, F., Najita, J., Ostriker, E., et al. 1994, ApJ, 429, 781
Shu, F. H., Shang, H., Gounelle, M., Glassgold, A. E., & Lee, T. 2001, ApJ, 548, 1029
Stępień, K., & Landstreet, J. D. 2002, A&A, 384, 554
Stelzer, B., Micela, G., Hamaguchi, K., & Schmitt, J. H. M. M. 2006, A&A, 457, 223
Stolker, T., Sitko, M., Lazareff, B., et al. 2017, ApJ, 849, 143
Szeifert, T., Hubrig, S., Schöller, M., et al. 2010, A&A, 509, L7
The, P. S., de Winter, D., & Perez, M. R. 1994, A&AS, 104, 315
Tout, C. A., & Pringle, J. E. 1995, MNRAS, 272, 528
Tutukov, A. V., & Fedorova, A. V. 2010, ARep, 54, 156

Vieira, S. L. A., Corradi, W. J. B., Alencar, S. H. P., et al. 2003, AJ, 126, 2971
Vink, J. S., Drew, J. E., Harries, T. J., & Oudmaijer, R. D. 2002, MNRAS, 337, 356
Vural, J., Kraus, S., Kreplin, A., et al. 2014, A&A, 569, A25
Wade, G. A., Drouin, D., Bagnulo, S., et al. 2005, A&A, 442, L31
Wade, G. A., Bagnulo, S., Drouin, D., Land street, J. D., & Monin, D. 2007, MNRAS, 376, 1145
Wheelwright, H. E., Oudmaijer, R. D., & Goodwin, S. P. 2010, MNRAS, 401, 1199
Wichittanakom, C., Oudmaijer, R. D., Fairlamb, J. R., et al. 2020, MNRAS, 493, 234

Swetlana Hubrig and Markus Schöller

Chapter 7

Other Objects

Over the last decade, observational evidence is accumulating that apart from magnetic Ap/Bp and O-type stars, also other non-degenerate stars with radiative envelopes can possess magnetic fields. HgMn, PGa, and Am stars show different abundance patterns compared to those observed in Ap/Bp stars and probably lack a large-scale organisation of their magnetic fields. The presence of small-scale variably structured magnetic fields was suggested for rapidly rotating hot Be stars with decretion disks. Slowly pulsating B and β Cephei stars seem to rarely harbour strong magnetic fields. The detection of magnetic fields in these stars is usually hampered by the presence of distinct spectral pulsational variability, which has an impact on the magnetic field measurements. Magnetic fields in the optical components of high-mass X-ray binaries and in the descendants of massive O-type stars, the Wolf–Rayet stars, are only rarely searched for, mostly due to their faintness.

7.1 HgMn Stars

Upper-main sequence B7–B9 HgMn stars exhibit in their atmospheres certain chemical abundance anomalies, i.e., large excesses of P, Mn, Ga, Br, Sr, Y, Zr, Rh, Pd, Xe, Pr, Yb, W, Re, Os, Pt, Au, and Hg and underabundances of He, Al, Zn, Ni, and Co (e.g., Castelli & Hubrig 2004a). Strong isotopic anomalies were detected for the chemical elements Ca, Pt, and Hg with patterns changing from one star to the next (Hubrig et al. 1999a; Castelli & Hubrig 2004b; Cowley et al. 2009). Observationally, these stars are characterised by low rotational velocities ($\langle v \sin i \rangle \leqslant 29$ km s^{-1}; Abt et al. 1972). Prior to the studies of Chojnowski et al. (2020) and Paunzen et al. (2021), about two hundred HgMn stars were known and many of them were found in young clusters and young associations like Sco-Cen, Orion OB1, or Aur OB1 (Renson & Manfroid 2009). According to North et al. (1997) and Hubrig et al. (2001, 2012), HgMn stars are uniformly distributed across the whole width of the main-sequence band. A number of HgMn stars are located very close to the zero-age main sequence (ZAMS).

doi:10.1088/2514-3433/abefccch7

Using SDSS/APOGEE H-band spectroscopy, Chojnowski et al. (2020) identified 269 HgMn stars among telluric standard stars based on their metallic absorption content being limited to or dominated by the H-band Mn II lines. Due to the fainter magnitudes probed by the APOGEE survey, in comparison with past studies, only nine out of the sample of 269 stars were previously known as HgMn stars. The 260 newly identified HgMn stars represent more than a doubling of the previously known sample of these stars. At least 32% of the APOGEE sample were found to be binary or multiple systems. Additional 99 HgMn stars, together with 19 good HgMn candidate stars and seven further candidates identified using spectra obtained with the Large Sky Area Multi-Object Fibre Spectroscopic Telescope (LAMOST) were reported by Paunzen et al. (2021). Using the colour–magnitude diagram based on homogeneous Gaia DR2 photometry from Arenou et al. (2018), the authors reported that the HgMn stars in their sample cover the whole age range from ZAMS to terminal-age main sequence, with an apparent accumulation of objects between fractional ages of 50%–80%.

More than two thirds of the HgMn stars are known to belong to spectroscopic binaries (Hubrig & Mathys 1995), with a preference of orbital periods in the range between 3 and 20 days. Orbital periods shorter than 3 days are common among normal late-B stars, but not among HgMn stars. About half of the HgMn binaries are SB2 systems. Many binaries with a HgMn primary were reported to rotate subsynchronously (e.g., Guthrie 1986; Hubrig & Mathys 1995). It is not clear whether subsynchronous rotation observed in these stars is related to the presence of a magnetic field or to the fact that numerous HgMn systems comprise more than two components. Among the systems definitely rotating subsynchronously, several systems showed the presence of distant companions (Hubrig & Mathys 1995).

It is striking that an inspection of SB systems with late B-type primaries in the 9th Catalogue of Spectroscopic Binary Orbits (Pourbaix et al. 2004) indicates a strong correlation between the HgMn peculiarity and the membership in a binary system: Among bright well studied SB systems with late B-type slowly rotating ($v \sin i < 70$ km s^{-1}) primaries with an apparent magnitude of up to $m_V \approx 7$ and orbital periods between 3 and 20 days, apart from HR 7241, all 21 systems have a primary with a HgMn peculiarity. Based on this fact, it is very likely that the majority of slowly rotating late B-type stars formed in binary systems with certain orbital parameters become HgMn stars. Therefore, careful studies of these peculiar stars are important for the general understanding of B-type star formation in binary systems. Since a number of HgMn stars in binary systems is found at the ZAMS (e.g., Nordstrom & Johansen 1994; González et al. 2010), it is expected that the timescale for developing HgMn peculiarities is very short.

According to speckle interferometric observations (Cole et al. 1992; Isobe 1991), diffraction-limited near-infrared imaging with NAOS-CONICA (NACO) at the VLT (Schöller et al. 2010) and observations of X-ray emission (Hubrig & Berghofer 1998; Hubrig et al. 2001), a large number of HgMn stars belong to triple or even quadruple systems. Out of 30 SB HgMn stars observed with speckle interferometry, 15 appeared to have more than two components. Schöller et al. (2010) used NACO in imaging mode with the S13 camera to observe 57 HgMn stars with K-magnitudes ranging from 3.7 to 9.1. The authors discovered visual companions down to $K = 18$ in the separation range 0.05–7.8″. 31 companions in 29 systems were detected, including three triple systems and one quadruple system. Among them, 17 SB systems have known orbital elements from the 9th catalogue of SB orbits (Pourbaix et al. 2004) and the Multiple Star Catalogue by Tokovinin (1997). Further evidence that HgMn stars frequently are members of multiple systems is inferred from the results of the ROSAT all-sky survey (Berghoefer et al. 1996). X-ray emission was detected through this survey in twelve HgMn stars (seven SB1s, three SB2s and two for which no radial velocity data are available). Previous X-ray observations with the Einstein Observatory and theoretical estimates had suggested that stars in the spectral range B2–A7 are devoid of any significant X-ray emission. In most cases where X-ray emission had been detected in such stars, it was found to originate from a cool companion. This suggests that the X-ray emission found in HgMn SBs does not originate from the HgMn primary (Hubrig & Berghofer 1998). Indirect evidence for the presence of a third component was found in four other HgMn SB systems (HD 11905, HD 34364, HD 78316, and HD 141556) on the basis of spectroscopic and photometric arguments. The high occurrence of HgMn stars in binary and multiple systems indicates that to understand star formation theories, it is important to compare the ranges of orbital periods, luminosity ratios, orbital eccentricities and hierarchy of HgMn stars with those for normal B stars.

7.1.1 Spectroscopic Variability

Previous spectroscopic studies revealed that line profile variability is a general characteristic of HgMn stars, rather than an exception (e.g., Nuñez et al. 2011). This variability is similar to the variability discovered in magnetic Ap/Bp stars and is caused by an inhomogeneous surface chemical element distribution. The number of stars with intrinsic spectroscopic and photometric variability is currently gradually increasing, implying that most HgMn stars present a non-uniform distribution of several chemical elements. The existence of an inhomogeneous distribution of some chemical elements over the surface of HgMn stars has been, for the first time, discussed by Hubrig & Mathys (1995). From a survey of HgMn stars in close SB systems, the authors suggested that some chemical elements might be concentrated in chemical spots, with, in particular, preferential concentration of Hg along the equator. In close SB2 systems, where the orbital plane has a small inclination to the line of sight, a rather large overabundance of Hg was found. By contrast, in stars with orbits almost perpendicular to the line of sight, Hg is not observed at all. The first definitively identified spectrum variability, which is not caused by a companion, was reported for the binary HgMn star α And by Wahlgren et al. (2001) and Adelman et al. (2002). They suggested that the spectral variations of the Hg II λ3984 line discovered in high-dispersion spectra are not due to the orbital motion of the companion, but produced by the combination of a 2.8 days rotation period of the primary and a non-uniform surface distribution of Hg concentrated in the equatorial region. The authors analysed the variability of the Hg II λ3984 line using a Doppler imaging (DI) code and detected high-contrast Hg spots located along the rotational equator.

Using multiple epoch high signal-to-noise (S/N) high-resolution spectroscopic observations of the eclipsing SB2 HgMn binary star AR Aur, it was demonstrated that the chemical spots detected in α And are not unique (Hubrig et al. 2006b). The ZAMS eclipsing binary AR Aur (HD 34364, B9V+B9.5V) with an orbital period of 4.13 days at an age of only 4×10^6 years belongs to the Aur OB1 association (Nordstrom & Johansen 1994). A third body in this system, with a minimum mass of 0.51 $M_\odot$ was discovered by Chochol et al. (1988). Due to the young age of AR Aur, it presents the best case for a study of evolutionary aspects of the chemical peculiarity phenomenon. The primary and the secondary eclipses in this system are nearly total since its orbital inclination is 88.5° and the radii of both stars are almost equal (Nordstrom & Johansen 1994). An analysis of the available light and radial velocity curves showed that the secondary star is still contracting toward the ZAMS, while the primary star appears to be exactly on the ZAMS.

The behaviour of the line profiles of a number of elements over the rotation period of the primary in the system AR Aur is shown in Figure 7.1. The Zr II, Nd III, Pt II, and He I lines in the presented plots appear rather weak, but still their variations are well notable. Interestingly, while the behaviour of the line profiles of Y II, Pt II, Hg II, Sr II, and Nd III is rather similar over the rotation period, the line profiles of Zr II and He I seem to vary with a 180° phase shift. The mean line profiles of Fe II obtained using six unblended lines are displayed in Figure 7.1(h). Although very weak distortions in the individual lines over the rotation period are detectable, the width of the mean line profile suggests that Fe is distributed almost homogeneously over the stellar surface. The same kind of slight distortion is also observed in

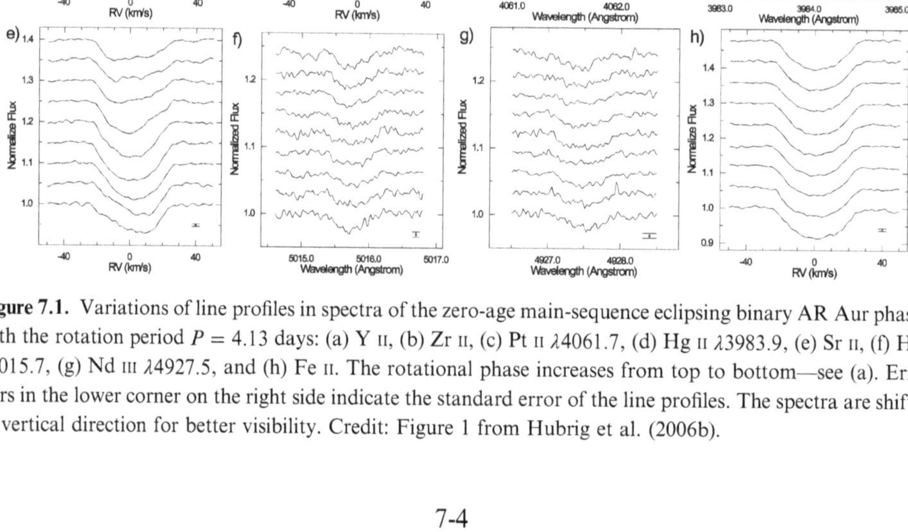

Figure 7.1. Variations of line profiles in spectra of the zero-age main-sequence eclipsing binary AR Aur phased with the rotation period $P = 4.13$ days: (a) Y II, (b) Zr II, (c) Pt II λ4061.7, (d) Hg II λ3983.9, (e) Sr II, (f) He I λ5015.7, (g) Nd III λ4927.5, and (h) Fe II. The rotational phase increases from top to bottom—see (a). Error bars in the lower corner on the right side indicate the standard error of the line profiles. The spectra are shifted in vertical direction for better visibility. Credit: Figure 1 from Hubrig et al. (2006b).

the O I, Na I, Mg II, Si II, Ca II, and Ti II line profiles. A comparison of the observed line profiles with synthetic spectra, calculated assuming the atmospheric parameters employed by Nordstrom & Johansen (1994), $T_{\rm eff}$ = 10 950 K and log g = 4.33, and assuming a microturbulent velocity $v_{\rm t}$ = 0.5 km s^{-1}, revealed that for these elements the overall shape of the line profiles slightly deviates from purely rotationally broadened profiles (Hubrig et al. 2006b).

As a result of subsequent extensive spectroscopic studies during the last years, it can be considered as established that chemical inhomogeneities are widespread among the HgMn stars and that previous failures to detect them were largely related to the small number of repeated observations of the same targets and the traditional focus on sharp-lined stars, for which the spectrum variations are much harder to detect. Hubrig et al. (2011a) obtained high-quality spectra of a representative sample containing several tens of HgMn stars using UVES at the VLT and the Fiberfed Extended Range Optical Spectrograph (FEROS) at the ESO 2.2-m telescope within the framework of ESO programs aimed at a careful study of line profile variations of various elements. About 2/3 of the sample showed distinct profile variability. Numerous examples of the variability of various spectral lines belonging to different elements in HgMn stars are presented in Figure 7.2.

The first preliminary DI models with the DI inversion code iAbu for the elements Mn, Sr, Y, and Hg using nine spectra of AR Aur observed with the UVES spectrograph in 2005 was presented by Savanov et al. (2009). In the more recent analysis of this system by Hubrig et al. (2010b) based on additional spectra obtained a few years later with the SOviet-FINnish (SOFIN) échelle spectrograph installed at the 2.56-m Nordic Optical Telescope on La Palma, the Coudé Spectrograph of the 2.0-m telescope of the Thüringer Landessternwarte and the SES spectrograph of the 1.2-m STELLA-I robotic telescope at the Teide Observatory, the improved DI code IA, introduced by Freyhammer et al. (2009), was used. This code employs Tikhonov regularisation in a way similar to the DI method described by Piskunov (2008) with a grid of 6° × 6°. The DI technique was applied after the spectrum of the secondary star was removed and after the orbital radial velocity was subtracted. The secondary in the AR Aur system did not show line profile variability and the luminosity-ratio between the components was constant, so methods for separating the spectra of the components (e.g., González & Levato 2006) worked appropriately.

The results of the applied DI technique to reconstruct the surface distributions of Fe, Y, and Sr using spectroscopic time series obtained in 2005 and in 2008/2009 are presented in Figures 7.3–7.5 with solar values corresponding to −4.50 dex for Fe, −9.79 dex for Y and −9.12 dex for Sr (Grevesse et al. 2010). The inspection of the resulting maps showed that Fe is overabundant by up to +1.5 dex and Y is overabundant by up to +3.9 dex in several spots, and Sr is overabundant up to +3.3 dex in the low latitude spot. The comparison of the DI map constructed for observations in 2005 (set1) and the map based on observations obtained in 2009 (set2) showed a remarkable evolution of the element distribution and overabundances. The positions and the shape of the spots with the highest Fe overabundance slightly changed from 2005 to 2009 and the level of the Fe overabundance showed a significant increase, especially in the spot located close to the equator at phases

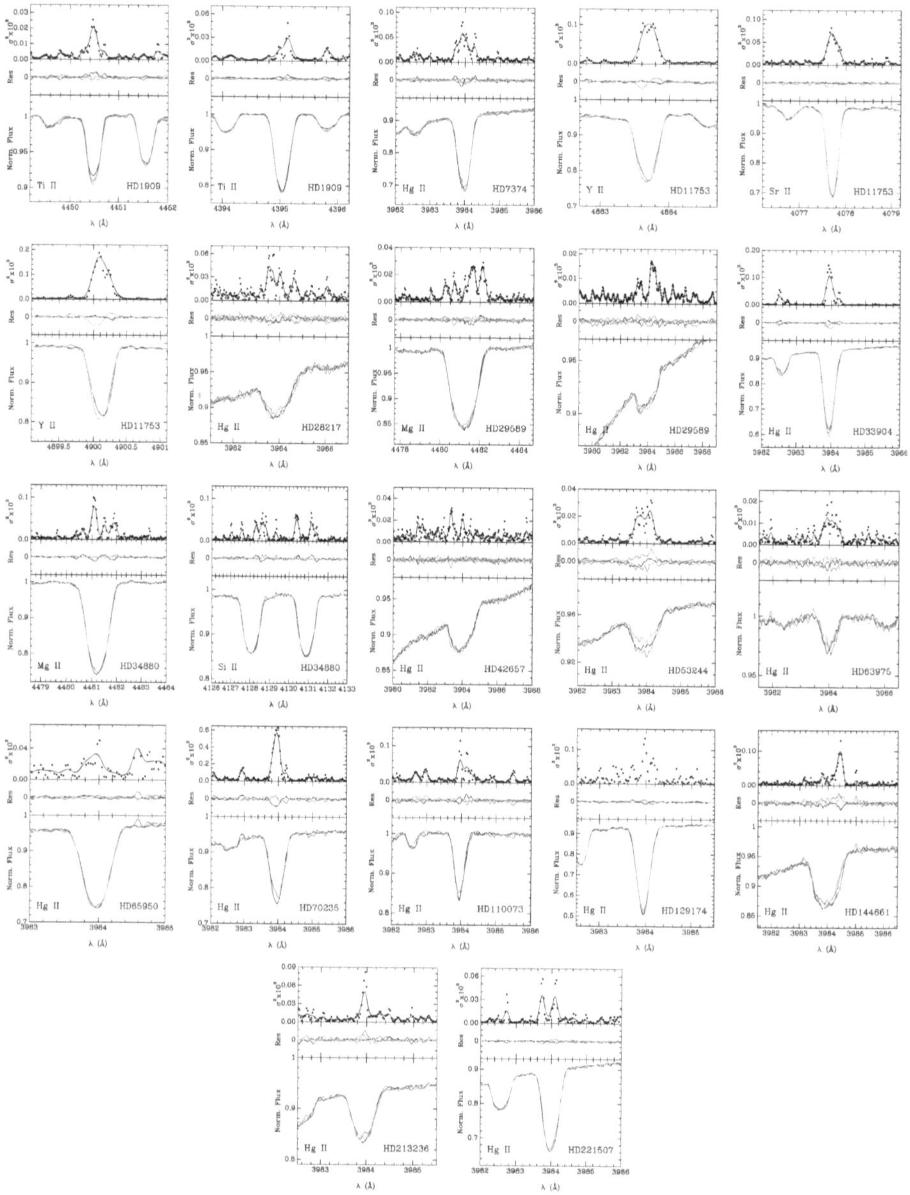

Figure 7.2. Examples of line profile variability in various HgMn stars. In each plot the middle panel has the same scale as the lower panel. The curves in the upper panels present a smoothing of the residual points using a convolution with a Gaussian of $\sigma = 2.05$ pixels. Credit: Figure 1 from Hubrig et al. (2011a).

0.50–0.75 and in the polar spot at phases 0.75–0.83. In the Y maps, the evolution of overabundance, shape and position of the spots appeared much more remarkable, revealing a region of huge overabundance having a shape of a belt that is broken around phase zero. Intriguingly, in this phase the hemisphere that is permanently

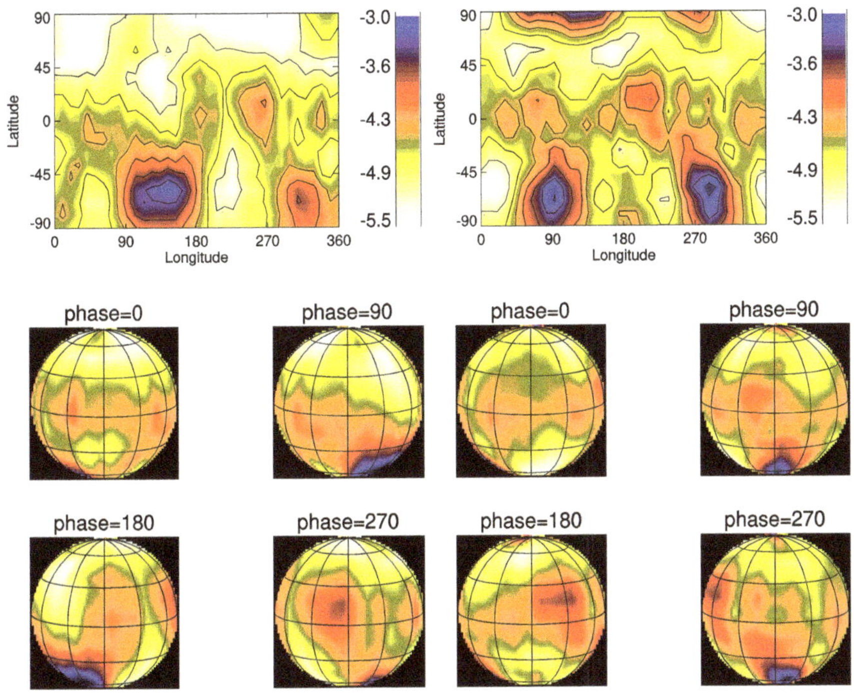

Figure 7.3. Fe abundance maps of AR Aur obtained from the Fe II λ4923.9 line for spectra from set1 (*left panels*) and set2 (*right panels*). The colour indicates the abundance with respect to the total number density of atoms and ions. Darker areas indicate higher element concentration. Credit: Part of Figure 1 from Hubrig et al. (2010b).

facing the secondary was observed. Such a behaviour was also observed for Sr, showing no overabundance in the same phase. The absence of Sr spots on the hemisphere that is permanently facing the secondary had already been discussed by Hubrig et al. (2006b). It is very likely that the lack of chemical spots on the hemispheres permanently facing the companion is caused by tidal interaction greatly affecting the upper atmospheric layers in stars that are members of close binary systems.

For one SB1 system with a well pronounced variability, the HgMn star HD 11753, it was possible to gather a large number of spectra with the CORALIE échelle spectrograph attached to the 1.2-m Leonard Euler telescope on La Silla in Chile (Briquet et al. 2010). In total, 113 spectra were obtained at a spectral resolution of 50,000. The radial velocities and equivalent widths in the spectra of HD 11753 were found to vary with the period $P = 9.54$ days. The DI technique was applied to reconstruct the surface distribution of Ti, Sr, and Y. Ti and Y have numerous transitions in the observed optical spectral region permitting to select unblended spectral lines with not too different line formation depths. The choice of lines with line formation depths is important to avoid discrepant abundance determinations due to possible elemental inhomogeneities in vertical direction.

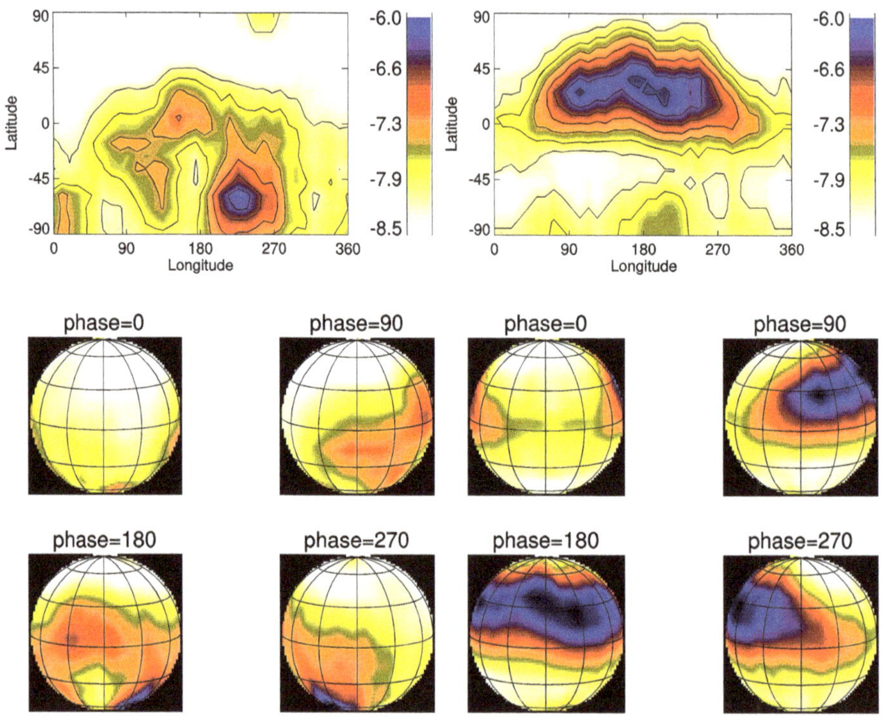

Figure 7.4. Y abundance maps of AR Aur obtained from the Y II λ4900.1 line for spectra from set1 (*left panels*) and set2 (*right panels*). Credit: Part of Figure 2 from Hubrig et al. (2010b).

The relevance of vertical abundance stratification in HgMn stars was previously discussed by Savanov & Hubrig (2003), who detected an increase in Cr abundance with height in the stellar atmosphere in a representative sample of stars with this peculiarity.

The two sets of observations of HD 11753 obtained in 2000 September (set1) and in 2000 December (set2) consisted of 76 and 28 observations, respectively, evenly spread over the stellar rotation cycle. Surprisingly, the results of the DI reconstruction revealed noticeable changes in the surface distributions of Ti, Sr, and Y between the data sets separated by just 65 days, indicating the presence of the hitherto not well understood physical processes in stars with radiative envelopes causing a rather fast dynamical chemical spot evolution, not of the order of years, but rather months. All Ti, Sr, and Y abundance maps revealed a structure reminiscent of broken rings of low and high abundance. As an example, the maps obtained from the Y II lines 4883 Å and 4900 Å are presented in Figure 7.6. The maps show broken abundance rings, with the high abundance region extending from latitude 45° to the pole. The Y abundance distribution shows a high latitude lower abundance spot around phases 0.2–0.4, similar to the Ti abundance maps. The equatorial region is dominated by a belt of lower abundance spots and the regions below the equator by high abundance features. The average abundance of the Y maps is −6.77, which is significantly higher than the solar abundance. Notably, all the features revealed in the maps show

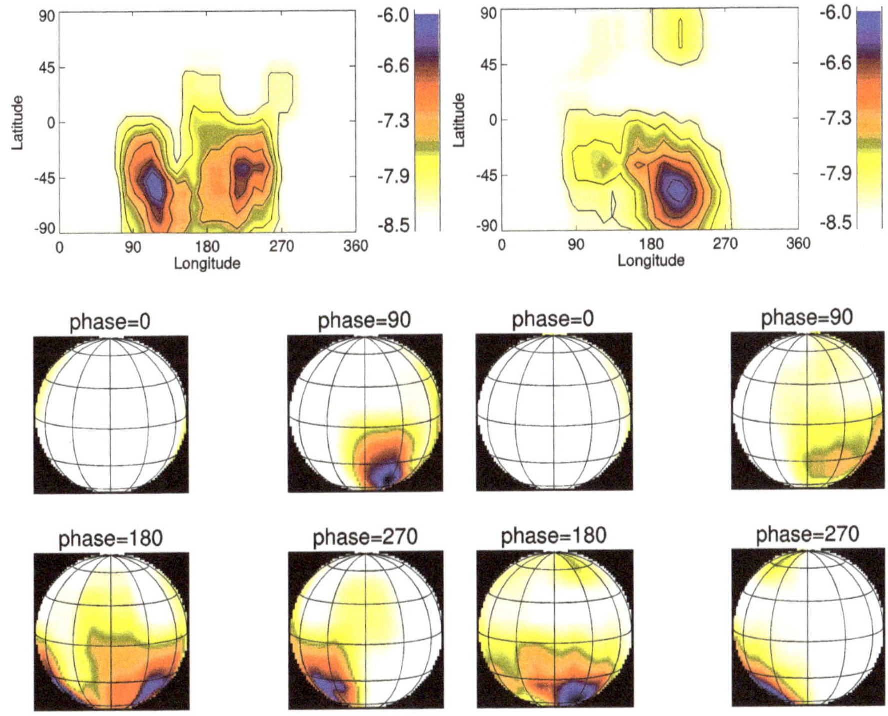

Figure 7.5. Sr abundance maps of AR Aur obtained from the Sr II λ4 215.5 line for spectra from set1 (*left panels*) and set2 (*right panels*).

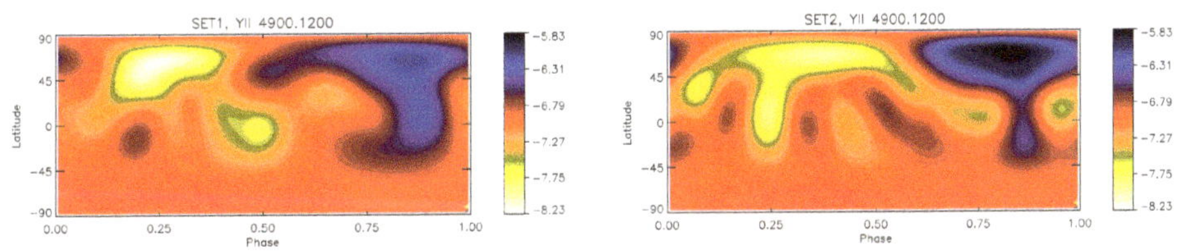

Figure 7.6. Y abundance maps of HD 11753 obtained for set1 (*left panel*) and set2 (*right panel*). The colour indicates the abundance with respect to the total number density of atoms and ions. Credit: Briquet et al. (2010), reproduced with permission © ESO.

abundances that are higher than the solar abundance of Y. Similar to the Ti maps, the Y maps show that the lower abundance high latitude feature at phases 0.2–0.4 becomes much less prominent in set2.

The abundance maps of HD 11753 presented in the work by Briquet et al. (2010) showed distinct differences in the spot configurations obtained from the same lines for the two data sets, indicating a rather fast dynamical evolution of the abundance distribution with time. Different dynamical processes take place in stellar radiation zones. An interaction between differential rotation, a magnetic field, and meridional circulation could possibly play a role in the generation of a dynamical evolution of the chemical spots. A comparison of the maps indicates that it is possible that the Y distribution shows indications of an increasing rotation rate toward the rotation pole, so-called differential rotation of anti-solar type. On the other hand, these results were not sufficient to claim the presence of a surface differential rotation. Further analyses of the element surface distribution in a larger sample of HgMn stars should be carried out before the implication of these results can be discussed in more detail.

Korhonen et al. (2013) confirmed the results of Briquet et al. (2010) for HD 11753 by an analysis using both the chemical spot maps and equivalent width measurements. The most prominent features in the Doppler maps remained at a similar location for the duration of the observations of about 10 years, but the exact shapes and abundances changed. In addition, the authors discovered a long-term decrease in the overall Y and Sr abundances. The detection of a secular and short-term dynamical evolution of the element spot distribution is of extreme importance to constrain theoretical studies of the mechanisms probably responsible for the surface structure formation in stellar atmospheres, such as magnetohydrodynamical instabilities, diffusion due to gravitational settling and radiative levitation or accretion of interstellar material in the presence of magnetic fields.

7.1.2 The Presence of Magnetic Fields

Typically, inhomogeneous chemical abundance distributions are observed only on the surface of magnetic chemically peculiar stars with large-scale organised magnetic fields. In these stars, the abundance distribution of certain elements is non-uniform and non-symmetric with respect to the rotation axis. Although strong large-scale magnetic fields have not generally been found in HgMn stars, it has never been ruled out though that these stars might have tangled magnetic fields of the order of a few thousand Gauss with no or very small net longitudinal components (Hubrig 1998). Highly structured magnetic field topologies can reduce the Stokes V signatures in spectral line profiles below the detection threshold. Furthermore, since HgMn stars usually display element spots on their surfaces, it is advisable to use in the magnetic field analysis the line masks for individual elements separately. Combining spectral lines of various elements may lead to the dilution of the magnetic signal or even to its cancellation, if enhancements of different elements occur in regions of opposite magnetic polarities.

Attempts to detect mean longitudinal magnetic fields in HgMn stars using low- and high-resolution spectropolarimetry have been made by several authors, who reached contradictory results. While low-resolution observations do not allow the use of individual elements separately, the analysis of high-resolution polarimetric spectra using the moment technique (Mathys 1993) applied to individual elements was indicative of the presence of weak longitudinal magnetic fields, crossovers and quadratic fields (e.g., Mathys & Hubrig 1995; Hubrig 1998; Hubrig et al. 2012). To achieve higher accuracies in the measurements of weak stellar magnetic fields, the least-squares deconvolution (LSD; Donati et al. 1997), combining hundreds and sometimes up to thousand lines of all elements was frequently applied. Using samples of lines belonging to different elements together, Makaganiuk et al. (2011)

were not able to obtain any definite detection of a mean longitudinal magnetic field in a sample of HgMn stars, but could decrease uncertainties down to below 10 G for the best cases. However, combining the lines of elements with different surface abundance distributions unavoidably leads to the dilution of the magnetic signal, which is expected to be rather weak in HgMn stars. Just in one case, the LSD study of the HgMn star HD 11753, did the authors use line lists for Ti, Y, and Cr separately (Makaganiuk et al. 2012). These measurements indicated a mean longitudinal magnetic field in the range from about −20 G to +25 G, with measurement uncertainties of up to 25 G.

Based on spectropolarimetric studies carried out during the last decades, it became clear that weak magnetic fields do exist in a number of HgMn stars (e.g., Hubrig et al. 2008b, and references therein). The most widespread method to detect a magnetic field is to obtain polarimetric spectra recorded in left- and right-hand polarised light to measure the mean longitudinal magnetic field. Another approach to study the presence of magnetic fields in upper-main sequence stars is to determine the value of the mean quadratic magnetic field $\langle B_{\rm q}\rangle = (\langle B^2\rangle + \langle B_{\rm z}^2\rangle)^{1/2}$ (see Section 2.8). The analysis is usually based on the consideration of samples of reasonably unblended lines of Fe I and Fe II. Applying this method, Mathys & Hubrig (1995) could demonstrate the presence of a quadratic magnetic field of the order of a few kG in the primary component of 74 Aqr using polarimetric spectra recorded on two consecutive nights.

The eclipsing binary AR Aur system was observed once in 2009 December at the rotational/orbital phase 0.622 with the SOFIN spectropolarimeter installed at the 2.56-m Nordic Optical Telescope on La Palma. A longitudinal magnetic field at a level higher than 3σ of the order of a few hundred Gauss was detected using Fe II, Ti II, and Y II lines, while a quadratic magnetic field of the order of several kG was measured using Ti II lines (Hubrig et al. 2010b). A weak longitudinal magnetic field of about 200 G was detected in the secondary component using a sample of nine Fe II lines. The diagnosis of the quadratic field is more difficult than that of the longitudinal magnetic field as it depends much more critically on the number of lines that can be employed. Interestingly, a weak longitudinal magnetic field was previously also detected in the secondary of another SB2 binary, χ Lup (Mathys & Hubrig 1995). The secondaries in both, the AR Aur and χ Lup systems, are mentioned in the literature to have characteristics very similar to early Am stars.

The only longitudinal magnetic field measurements carried out for AR Aur were reported by Folsom et al. (2010), who used ESPaDOnS and Narval observations obtained in 2006 on six epochs and applied the LSD technique to a sample of 1168 lines belonging to different elements. No magnetic field was detected at any epoch in the analysis of the polarimetric spectra. As discussed above, combining the lines of elements with different surface abundance distributions can lead to the dilution of the magnetic signal or even to its cancellation. Since the authors reported that no variability of the Ti and Fe lines was detected in their spectra, another possibility for the non-detection of the longitudinal magnetic field can be related to an unfavourable element spot configuration or even to the absence of some element spots at the epoch of their observations obtained in 2006. The surface distribution of Ti, Fe,

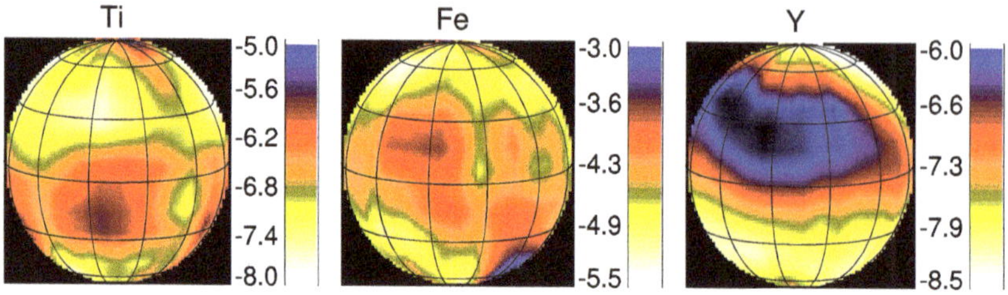

Figure 7.7. The element distribution from the Ti II λ4563.8 line, the Fe II λ4923.9 line and the Y II λ4900.1 line at the time of the magnetic field measurement (phase 0.622) for AR Aur. Credit: Figure 5 from Hubrig et al. (2010b).

and Y at the rotational/orbital phase 0.622 obtained using observations in 2008/2009 (set2) is presented in Figure 7.7. The observations used to map the distributions of these elements were acquired at a time closest to the SOFIN spectropolarimetric observation in 2009. Obviously, a higher Fe, Ti, and Y concentration in the surface spots is well detectable at this epoch corresponding to the phase 0.622. Notably, the distribution of Ti, Fe, and Y in the map based on observations obtained in 2005 (set1) did not show a strong spot concentration at the same rotational/orbital phase (Hubrig et al. 2010b).

Another method that has been applied to study magnetic fields in HgMn stars employed the relative magnetic intensification of the two Fe II lines of multiplet 74 at λ6147.7 and λ6149.2, which have the same equivalent width to within 2.5% in non-magnetic late B- and early A-type stars, but very different Zeeman patterns. This method was previously introduced by Mathys & Lanz (1990) and widely used in other studies (e.g., Hubrig et al. 1999b; Hubrig & Castelli 2001). The observed discrepancy between the equivalent widths of these two lines in magnetic stars is then attributed to magnetic intensification. The relative magnetic intensification of these two lines, roughly correlating with the strength of the magnetic field, appears to be a powerful tool for detecting stellar magnetic fields that have a complex structure and are difficult to analyse by polarisation measurements. Evidence for the presence of relative magnetic intensification of the Fe II lines of multiplet 74, produced by different magnetic desaturations induced by different Zeeman-split components, was previously shown for a few HgMn stars by Hubrig & Castelli (2001).

It was previously argued that magnetic fields of Ap stars may be generated by strong binary interaction, i.e., in stellar mergers or during a mass transfer or common envelope evolution (e.g., Tout et al. 2008). Two more scenarios for the origin of the magnetic fields of Ap stars, in which these stars result from the merging of two lower mass stars or protostars, were suggested by Tutukov & Fedorova (2010) and Ferrario et al. (2009), respectively. If the merger scenario is valid, one should expect a very small number of Ap stars to be members of close binaries and these binaries could probably be considered as failed merger products. In contrast to HgMn stars, which are frequently members of close double-lined systems, close binaries with one or two classical magnetic Ap/Bp components are very rare. Among the studied binary systems with A-type components, only two systems,

HD 98088 and HD 161701, were known to have as a component a magnetic Ap star. The Ap component in the system HD 98088 with $P_{orb} = 5.9$ days exhibits a primary intensity maximum of the rare earth element europium near the phase of superior conjunction, i.e., at the surface persistently facing the companion classified as Am star (Abt et al. 1968). This behaviour is similar to that found in the Ap component in the binary system HD 161701 with $P_{orb} = 12.45$ days, which consists of a HgMn primary and a magnetic Ap secondary (Hubrig et al. 2014a). Intriguingly, the magnetic dipoles of the Ap components in both systems are aligned along the axis between the two stars, with the positive magnetic pole always pointing to the secondary. To demonstrate this behaviour in HD 161701, the SVD Stokes *I*, *V* and diagnostic null profiles obtained with HARPSpol over six consecutive nights using Fe II and Fe I lines are shown in Figure 7.8, where the clear Zeeman features correspond to the position of the Ap component. According to studies of magnetic instabilities (e.g., Arlt & Rüdiger 2011), the alignment of the magnetic axis with the orbital radius vector may indicate that the generation of the magnetic field was a dynamic process during tidal synchronisation, i.e., tidal forces would alter the flows during the unstable phase to align the final field geometries in the observed way. To properly evaluate theoretical models of the origin of magnetic fields in Ap/Bp stars and to investigate whether magnetic fields of these stars are tidally trapped in close binaries, further systematic studies of the magnetic field configuration and spectral variability in close binaries should be conducted.

Future studies of close binaries with peculiar components and weak magnetic fields are also needed to probe the correlations between binary properties, magnetic field structure and abundance inhomogeneities. From a comparison of the physical properties of the binary components, which formed from the same original material

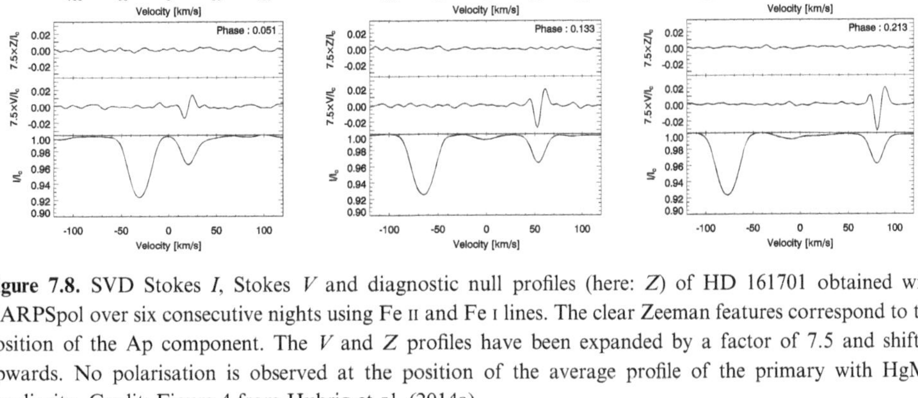

Figure 7.8. SVD Stokes *I*, Stokes *V* and diagnostic null profiles (here: *Z*) of HD 161701 obtained with HARPSpol over six consecutive nights using Fe II and Fe I lines. The clear Zeeman features correspond to the position of the Ap component. The *V* and *Z* profiles have been expanded by a factor of 7.5 and shifted upwards. No polarisation is observed at the position of the average profile of the primary with HgMn peculiarity. Credit: Figure 4 from Hubrig et al. (2014a).

and have evolved in the same environment, one should be able to gain more insight into the mechanisms that are responsible for the development of chemical peculiarities and the generation of magnetic fields. Since classical Ap stars with large-scale organised kG magnetic fields are extremely rare members of close binary systems, this information is also critical for answering the principal question of the possible origin of magnetic fields in these stars.

A highly peculiar chemical composition of the late-B HgMn primary in the eclipsing system HD 66051 with $P_{\rm orb} = 4.75$ days was announced by Niemczura et al. (2017). Besides overabundances of the elements Mn, Xe, Hg, Pt, and Pb, typical for the chemical composition of HgMn stars, also rare earth element overabundances typical for magnetic Bp stars were detected (Niemczura et al. 2020). The analysis of the secondary indicated an overabundance of iron-peak, heavy and rare earth elements. Using observations with the ESPaDOnS spectropolarimeter at the Canada–France–Hawaii Telescope (CFHT), Kochukhov et al. (2018) reported the presence of a weak magnetic field of just a few tens of Gauss in the primary and concluded that the primary is not a HgMn star, but a classical magnetic Bp star. This classification is certainly doubtful, as overabundances of Xe and the heavy elements Hg, Pt, and Pb were never detected in magnetic Bp stars. In contrast, a moderate rare earth element overabundance was reported in the past for numerous HgMn stars (e.g., Castelli & Hubrig 2004a; González et al. 2014).

A previous survey of magnetic fields in 17 HgMn stars (Hubrig et al. 2006c) using low-resolution ($R = 2000$) circular polarisation spectra obtained with FORS 1 at the VLT led to field detections in four stars. One of them, the HgMn star HD 65949 with a detected weak magnetic field, was followed-up by additional FORS 1/2 observations in 2007 and in 2011. The results of these observations confirmed the presence of the magnetic field with $\langle B_z \rangle = -116 \pm 32$ G measured in 2007 and $\langle B_z \rangle = -182 \pm 34$ G measured in 2011 (Hubrig et al. 2012). As reported by Cowley et al. (2010), HD 65949 belongs to the young open cluster NGC 2516 and is a SB1 system with an orbital period of 21.2 days. Interestingly, the authors reported on a small variation in the centre-of-mass velocity, which was interpreted as due to the presence of a third body, making HD 65949 a triple system. HD 65949 exhibits enormous enhancements of heavy elements, in particular of the elements rhenium and osmium through mercury ($Z = 75 - 80$). According to Cowley et al. (2010), osmium and rhenium are rarely enhanced to the point where they are identified in ground-based spectra of HgMn stars. The authors suggested that the high abundance in HD 65949 must result from a very recent transfer of material.

An analysis of a few high-resolution HARPSpol observations of HD 65949 revealed the presence of a distinct spectral line variability (Hubrig et al. 2020a). Furthermore, the application of the LSD technique to the HARPSpol spectra indicated the possible presence of local tangled magnetic fields, which are sampled differently by different elements. A rather strong longitudinal magnetic field up to $\langle B_z \rangle = 177 \pm 28$ G with FAP $< 2.7 \times 10^{-8}$ was measured using He and Ti lines and up to $\langle B_z \rangle = -126 \pm 19$ G with FAP $< 8.2 \times 10^{-6}$ was measured using Cr and Fe lines (Hubrig et al. 2020a) in HARPSpol spectra obtained in 2019 June. The detection of a mean longitudinal magnetic field in this star is in agreement with the previous detections reported by Hubrig et al. (2006c, 2012), who used FORS 1/2 in spectropolarimetric mode installed at the ESO/VLT. The resulting LSD profiles for the HARPSpol spectra are shown in Figure 7.9. The positive polarity of the longitudinal magnetic field $\langle B_z \rangle$ measured using the sample of He and Ti lines and the negative polarity measured using the sample of Cr and Fe lines suggest that small-scale localised magnetic fields associated with chemical spots probably exist on the surface of HgMn stars.

The HgMn star HD 221507 was observed with HARPSpol on ten different epochs from 2010 to 2019. The results of the LSD technique applied to these spectra are presented in Figure 7.10. A positive mean longitudinal magnetic field $\langle B_z \rangle = 30 \pm 5$ G with FAP $< 2.4 \times 10^{-6}$ was definitely detected using Y lines for the spectra acquired in 2010. At the same epoch, only a marginal detection of a field of negative polarity $\langle B_z \rangle = -13 \pm 4$ G with FAP $< 6.4 \times 10^{-5}$ was achieved using the sample of Mn lines. Hubrig et al. (2012) applied the moment technique to a sample of Y lines

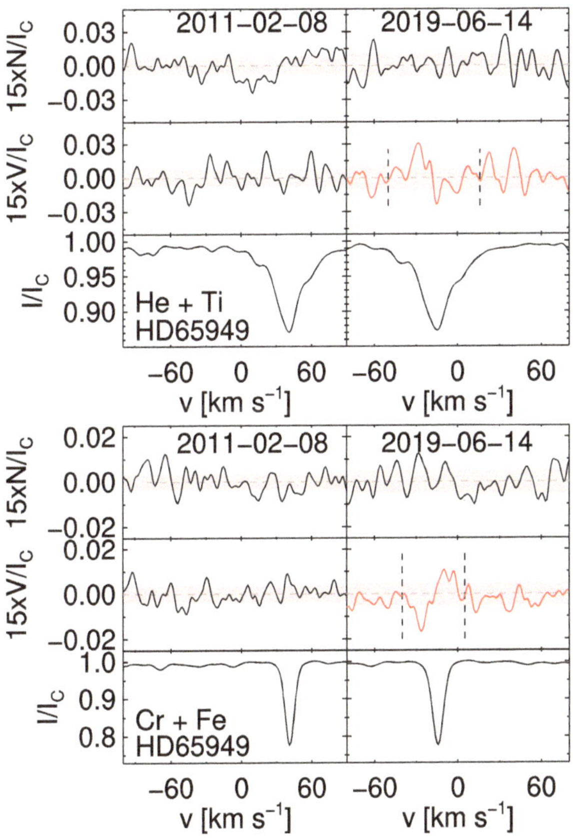

Figure 7.9. LSD Stokes *I* (*bottom*), Stokes *V* (*middle*), and diagnostic null (N) profiles (*top*) for HD 65949 observed on two different epochs with HARPS. The upper panel shows the results of the analysis using He and Ti lines and the lower panel the analysis using Cr and Fe lines. Credit: Part of Figure 2 from Hubrig et al. (2020a).

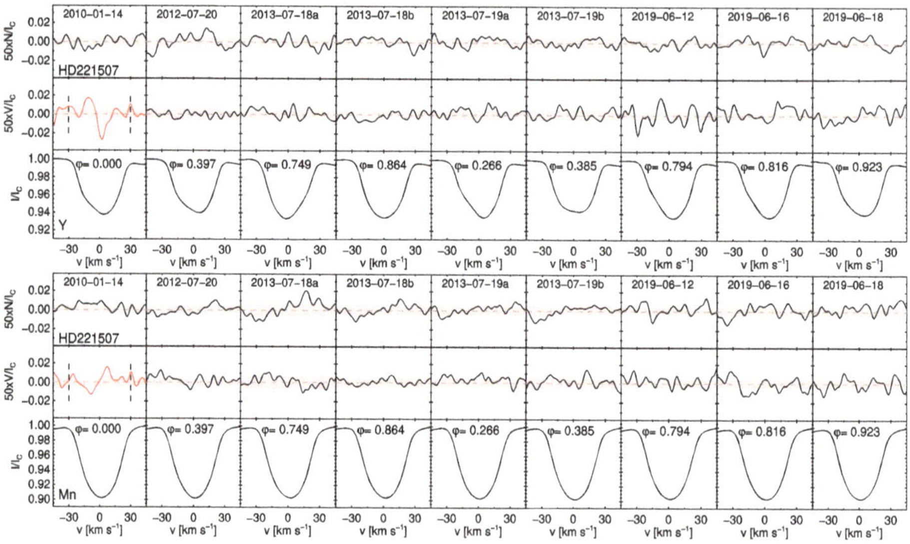

Figure 7.10. LSD Stokes *I* (*bottom*), Stokes *V* (*middle*), and diagnostic null (N) profiles (*top*) for HD 221507 using Y lines (*upper panel*) and Mn lines (*bottom panel*). The horizontal dashed lines indicate the $\pm 1\sigma$ ranges. Credit: Figure 1 from Hubrig et al. (2020a).

observed in the same HARPSpol spectra and reported $\langle B_z \rangle = 78 \pm 25$ G, which is in agreement with the presented LSD measurements for this element, $\langle B_z \rangle = 30 \pm 5$ G, within the error bars.

In the same study, Hubrig et al. (2020a) discussed two HARPS spectropolarimetric observations of the Y and Zr rich HgMn star HD 101189, which is not known to belong to an SB system, but seems to be a member of a visual binary. Schöller et al. (2010) detected a close companion candidate at a separation of 0.337″ and a position angle of 104.1° using adaptive optics near-infrared observations. As is shown in Figure 7.11, HD 101189 appears to be a good representative of typical HgMn stars, exhibiting a distinct variability of line profiles of several elements. However, only marginal field detections of the order of a few Gauss were achieved for this star.

The opposite polarities of the longitudinal magnetic field $\langle B_z \rangle$ measured using different line samples indicates the possible presence of local tangled magnetic fields, which are sampled differently by different elements. A number of studies using the moment technique already indicated the presence of tangled magnetic fields in a few HgMn stars (e.g., Hubrig et al. 2012). However, as the measured longitudinal magnetic field strengths in these stars were rather low, the suggestion of the presence of complex fields was usually based on the discovery of quadratic magnetic fields. This was the first time that the analysis of longitudinal magnetic fields showed a hint of evidence for a tangled structure of the surface magnetic field in HgMn stars (Hubrig et al. 2020a). The presence of opposite field polarities in temperature spots in the late-type star II Peg was recently discussed by Strassmeier et al. (2019) and appears to be typical in magnetically active late-type stars. However, the detection of

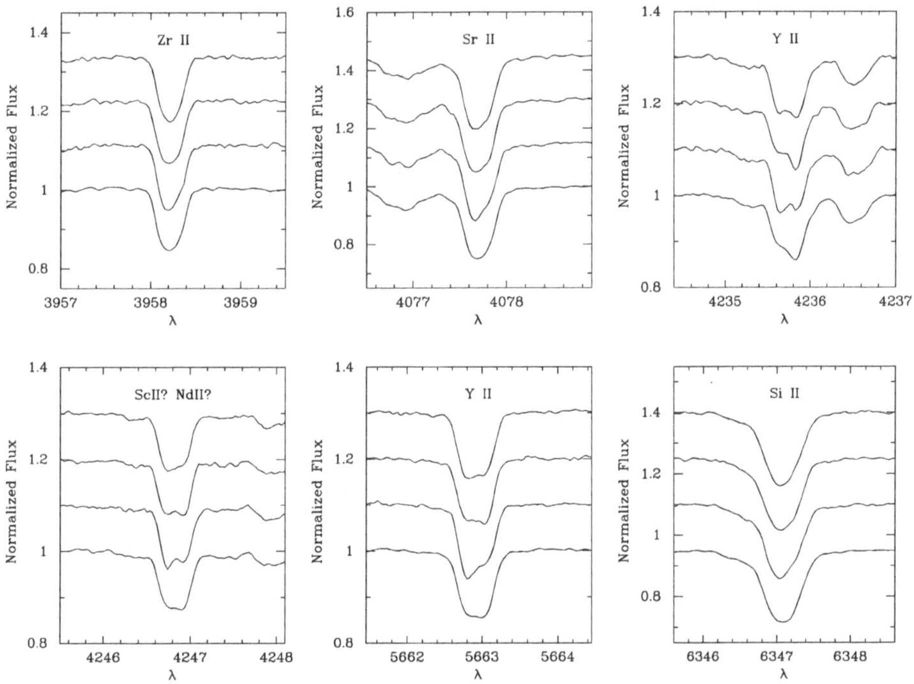

Figure 7.11. Various degrees of variability of line profiles in the spectra of HD 101189 obtained with FEROS on four different nights. Individual spectra are shifted in vertical direction for better visibility. Credit: Figure 7 from Hubrig et al. (2011a).

opposite field polarities using different chemical elements is completely unexpected in stars with radiative envelopes.

Since magnetic fields were not discovered in all of the available HARPSpol observations obtained at different epochs distributed over several years, it is quite possible that their detectability strongly depends on the spot position and the element abundance within the spot, i.e., on the "evolutionary states" of the spots especially when observing epochs are separated by long time intervals. As discussed above, noticeable long- and short-term changes in the surface element distribution take place in the atmospheres of HgMn stars. This means that a single spectropolarimetric observation of a HgMn star is not sufficient to judge whether it is magnetic or not. Following the discovery of tangled weak magnetic fields in HD 65949 and HD 221507, an analogous search for magnetic fields needs to be initiated for a representative sample of HgMn stars.

7.1.3 Scenarios for Weak Magnetic Field Generation in HgMn Stars

A scenario how a magnetic field can be built up in binary systems has been presented by Hubrig (1998). Tidal despinning to synchronous rotation proceeds from the outside toward the inside of the star. It can be assumed that the rotation at the surface of HgMn stars in binary systems is a poor indicator of the total angular momentum stored in these stars and that the angular velocity is decreasing outwards. The tidal torque per unit mass varies with depth and latitude in a star and therefore tends to induce differential rotation (Zahn 1984). Differential rotation in a radiative star can be prone to magneto-rotational instability (MRI; Arlt et al. 2003). Recent magneto-hydrodynamical (MHD) simulations revealed a distinct structure for the magnetic field topology similar to the fractured elemental rings observed on the surface of HgMn stars (Hubrig et al. 2008b). MHD simulations for magnetic field and velocity on the surface of a radiative star during the onset of the magneto-rotational instability are presented in Figure 7.12. The flows and fields resulting from the instability efficiently redistribute angular momentum and deliver a uniformly rotating star after 10–100 Myr. The magnetohydrodynamical simulations reveal a distinct structure for the magnetic field topology somewhat similar to the fractured element rings observed on the surface of HgMn stars. The initial model differential rotation was hydro-dynamically stable (Taylor–Proudman flow), but the introduction of a magnetic field excites the MRI on a very short timescale compared to the timescale of microscopic magnetic diffusion. Although the fields are not very strong, complex surface patterns can be obtained from the nonlinear, non-axisymmetric evolution of the MRI as shown in Figure 7.13 (Hubrig et al. 2011a).

It is important that future research projects concentrate on the search for a link between the magnetic nature and the fundamental properties of a representative sample of spectroscopic binaries with late B-type primaries using spectroscopic and spectropolarimetric observations and the magnetic Doppler imaging technique, which should be complemented by numerical simulations of magnetic stellar configurations. The discovery of exotic abundances, chemical inhomogeneities and weak magnetic fields on the surface of late B-type primaries in spectroscopic binaries

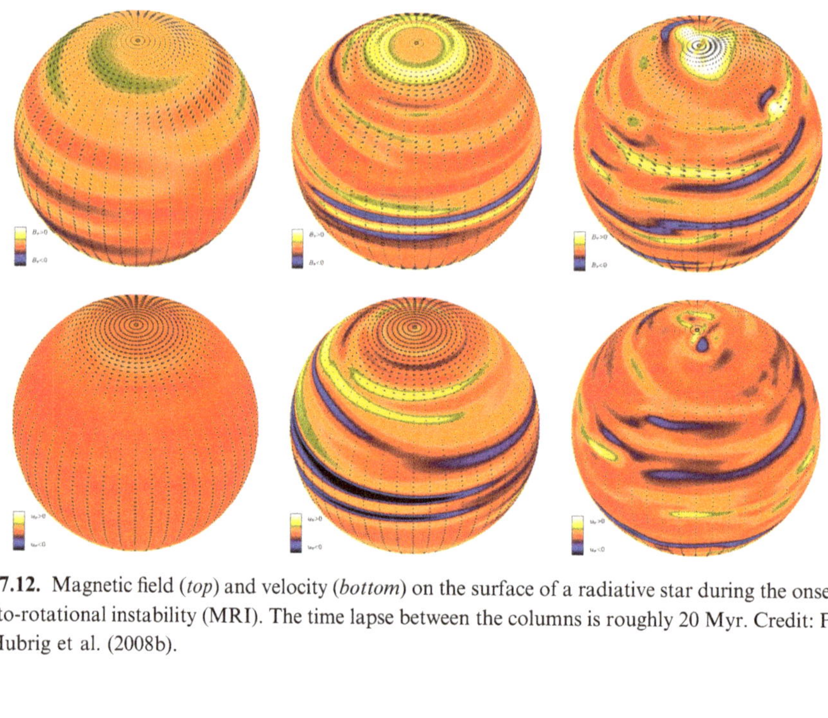

Figure 7.12. Magnetic field (*top*) and velocity (*bottom*) on the surface of a radiative star during the onset of the magneto-rotational instability (MRI). The time lapse between the columns is roughly 20 Myr. Credit: Figure 1 from Hubrig et al. (2008b).

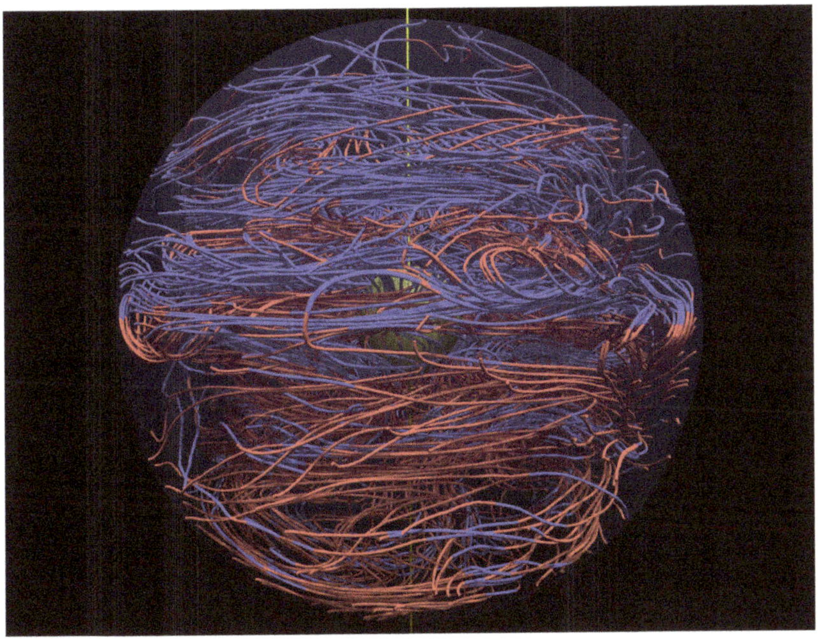

Figure 7.13. Magnetohydrodynamic simulations reveal a distinct magnetic field topology, which is similar to the fractured element abundance rings observed on the surface of several HgMn stars using DI. Credit: Figure 9 from Hubrig et al. (2011a).

has important implications not only for the understanding of the formation mechanisms of stars with Hg and Mn peculiarities themselves, but also to the general understanding of B-type star formation.

7.2 PGa Stars

PGa stars are found between spectral types B6 and B4 and are characterised by a deficiency of He and overabundances of P and Ga in their atmospheres. These stars with effective temperatures of about 13 500 K and higher (e.g., Alonso et al. 2003; Rachkovskaya et al. 2006) can be considered as the hotter analogues of HgMn stars (e.g., Borra et al. 1983), as they frequently show overabundance of Mn and Hg. Some stars show line profile variability of Hg, P, Mn, Fe, and Ga. It was suggested that the variability of the line profiles of these elements is caused by their non-uniform surface distribution, similar to the presence of chemical spots detected in HgMn stars (Hubrig et al. 2014b). Similar to HgMn stars, PGa stars do not possess strong large-scale organised magnetic fields, but more complex fields of kG order cannot be definitely excluded (Hubrig et al. 2006c).

The measurement results of the typical PGa star HD 19400 using high-resolution spectropolarimetry are indicative of the potential presence of a weak magnetic field in HD 19400 (Hubrig et al. 2014b). Applying the moment technique to Mn lines, Hubrig et al. (2014b) measured a weak negative longitudinal magnetic field $\langle B_z \rangle = -70 \pm 23$ G at a significance level of 3σ on one epoch. The results from the SVD analysis of the observations at the same epoch using Mn II lines showed the longitudinal magnetic field $\langle B_z \rangle = -76 \pm 25$ G and for the SVD analysis of Fe II lines $\langle B_z \rangle = -91 \pm 35$ G. However, the obtained FAP values, 0.008 and 0.003, respectively, were above the value 10^{-3} and thus too high for a marginal detection according to Donati et al. (1992). The work of Hubrig et al. (2014b) was based on spectra obtained only on three different nights and as of today, HD 19400 is the only PGa star for which a search for the presence of a magnetic field and of an inhomogeneous element distribution has been carried out using high-resolution spectropolarimetry. Owing to the small number of detailed high-resolution, high S/N spectropolarimetric studies of PGa stars, the link between the chemically peculiar groups of the mid to late main sequence B stars, the magnetic Bp stars, the HgMn stars and the PGa stars, is not understood yet. To get a better insight into the nature of the PGa stars, future studies based on spectropolarimetric monitoring over their rotation periods and involving a representative sample of PGa stars are needed.

7.3 Am Stars

Chemically peculiar metallic-line stars, the so-called Am stars, have spectral types between early-F and early-A and show moderately underabundant Ca and Sc, while iron-peak elements, Sr, Y, Zr, Ba, and rare earth elements are enhanced (e.g., Titus & Morgan 1940; Roman et al. 1948; Conti 1965). Am stars are slow rotators compared to the normal A or F type stars, with $v \sin i$ usually less than 120 km s^{-1} (Abt & Moyd 1973). The reason for their slow rotation can be related to the fact that almost all Am stars are members of binary or triple systems (Abt 1961; Carquillat & Prieur 2007; Hubrig et al. 2010a). For the majority of the systems with Am components, the orbital periods are between 1.2 and 100 days.

To explain the origin of the abundance anomalies in chemically peculiar stars, radiative diffusion is frequently invoked (Michaud 1970) and is expected to cause the Am phenomenon even if a magnetic field is absent (Watson 1971). The slow rotation further assists the diffusion to segregate the elements (Michaud et al. 1983). Also accretion processes may strongly influence the surface abundances, such as the mass transfer from an evolved companion (Fowler et al. 1965; Proffitt & Michaud 1989).

Am stars do not possess large-scale organised kG magnetic fields similar to those detected in Ap stars, but it cannot be ruled out that they may have fields of a kG order with more complex configurations (Lanz & Mathys 1993; Savanov 1994). Weak but still detectable magnetic fields were measured in a few Am stars (e.g., Sirius A—Petit et al. 2011; β UMa and θ Leo—Blazère et al. 2016). As strongly magnetic stars are rarely found in spectroscopic binaries, earlier explanations involved the idea that strong magnetic fields prevent close binary formation. However, it is possible that binarity is affecting magnetism and not the other way around (Budaj 1999; see also Hubrig 1998 and Hubrig & Mathys 1995).

7.4 Normal and Superficially Normal A and B Stars

Lignières et al. (2009) and Petit et al. (2010) detected a magnetic field strength of order of one Gauss in Vega. The detection of a magnetic field in Vega suggests that the number of stars with weak surface fields might be quite large. Hence it is necessary in future surveys to obtain spectropolarimetric observations with very high S/N to decrease the uncertainties of the magnetic field measurements.

A few normal and superficially normal A and B stars were included in magnetic field studies of Ap/Bp stars as comparison stars. Seven non-pulsating normal early- to mid-B stars were observed with FORS 1 at the VLT along with a sample of β Cephei and slowly pulsating B stars. A weak magnetic field was detected in the early B-type star HD 52089 and in the B5IV star HD 153716 (Hubrig et al. 2009a). The presence of a weak magnetic field in the star HD 52089 was confirmed by subsequent observations using high-resolution HARPSpol spectra (Fossati et al. 2015).

The magnetic fields of two further stars, HD 179761, classified as a normal B-type star, and HD 209459, classified as a superficially normal B-type star, were studied using high and low-resolution spectropolarimetric observations (e.g., Hubrig et al. 2012). Superficially normal B-type stars are generally indistinguishable from normal stars in terms of their iron-peak element abundances (e.g., Smith & Dworetsky 1993). They are usually sharp-lined stars that possess normal MK classifications, but nonetheless exhibit mild peculiarities when observed at high spectral resolution. The HARPS spectropolarimetric material for HD 209459 revealed the absence of Hg and Y lines and the presence of weak Mn lines. The lines of Ti, Cr, and Fe appeared rather strong and showed a weak variability in the spectra of HD 209459, whereas the Ti, Cr, Mn, and Y lines appeared weak in the spectra of HD 179761 (Hubrig et al. 2012). Hubrig et al. (1999b) and Hubrig & Castelli (2001) studied the presence of a magnetic field in both stars using an anomalous strength of the Fe II λ6147.7 line relative to the Fe II λ6149.2 line. In this method, previously introduced by Mathys & Lanz (1990), the observed relative differences between the equivalent widths of the two Fe II lines are compared with those derived from synthetic spectra computed by neglecting magnetic field effects. While the differences between the equivalent widths for HD 209459 were found to lie within the error limits, the differences for HD 179761 were found to be much larger, even if all observational and computational uncertainties were taken into account (Hubrig & Castelli 2001). The detected relative differences in intensities between the two Fe II lines in the spectra of HD 179761 suggested the presence of a magnetic field. Both HD 179761 and HD 209459 were also observed with FORS 1 at the VLT by Hubrig et al. (2006c), who reported a 3σ detection for the mean longitudinal magnetic field of HD 179761. Observations with HARPSpol indicated the presence of a weak longitudinal magnetic field in HD 209459 at one epoch, measured using the samples of Ti, Cr, and Fe lines. A quadratic magnetic field of the order of 3 kG at a 3σ significance level was detected in HD 179761 using Fe lines and much weaker quadratic fields of the order of 1.3–1.5 kG were detected in observations of HD 209459 using Cr lines (Hubrig et al. 2012). The presence of a ~3 kG quadratic field in HD 179761 agrees well with the previous analysis of the relative magnetic intensification of the two Fe II lines by Hubrig & Castelli (2001).

It was not the first time that quadratic magnetic fields were detected in upper-main sequence stars considered as normal or superficially normal stars. Mathys & Hubrig (2006) studied the presence of a quadratic magnetic field in the superficially normal A-type star HD 91375. Their analysis led to the detection of a 1.7 kG quadratic field at a significance level of more than 4σ. As a matter of fact, HD 91375 was not part of the sample studied by these authors. One of the motivations for including HD 91375 in the study was to use it as a magnetically null check star, for verification that the employed analysis method did not yield spurious magnetic field detections.

Although only rather weak longitudinal magnetic fields are detected on the surface of normal and superficially normal stars, the detections of mean quadratic magnetic fields strongly suggest that magnetic fields are present in their atmospheres. Moreover, there is an obvious advantage of carrying out quadratic magnetic field studies, as they provide measurements of the magnetic field strengths that are less sensitive to the magnetic field structure and especially suited for magnetic fields with complex structure.

7.5 Slowly Pulsating B and β Cephei Stars

Recent high-precision uninterrupted high-cadence space photometry using a number of satellites (e.g., WIRE, MOST, CoRoT, Kepler, BRITE, TESS) led to a revolutionary change in the observational evaluation of the variability of massive stars. Based on the results of photometric monitoring using various satellites, it became clear that a large fraction of massive stars show photometric variability due to either β Cephei- or SPB-like pulsations, stochastic p modes or convectively-driven internal gravity waves.

β Cephei stars are main-sequence short-period (3–8 h) pulsating variables of spectral type O9 to B3, corresponding to a mass range of 8–20 $M_{\odot}$, that pulsate in low-order pressure (p) and/or gravity (g) modes. SPB stars show variability with periods approximately half a day and a few days, are less massive (3–9 $M_{\odot}$) main-sequence B-type stars and have multiperiodic high-order low-degree g mode oscillations. SPB stars are located in the H-R diagram just below the β Cephei stars.

Neiner et al. (2003a) presented the first detection of a magnetic field in the SPB star ζ Cas, whose dominant pulsation mode has a period $P = 1.56$ days. Two other detections of a magnetic field in pulsating B stars were achieved for β Cephei stars: β Cep (Henrichs et al. 2000) and V2052 Oph (Neiner et al. 2003b). The measured mean longitudinal magnetic fields in all three stars were rather weak, less than 100 G.

A long-term monitoring project aimed at asteroseismology of a large sample of slowly pulsating B (SPB) stars and β Cephei stars was started by researchers of the Institute of Astronomy of the University of Leuven about twenty years ago. Together with astronomers working at ESO, they performed between 2003 to 2005 the first systematic search for magnetic fields in SPB and β Cephei stars, using FORS 1 at the VLT. In their first publication on the results of the magnetic survey of eight β Cephei and 26 SPB stars, based on more than 80 FORS 1 spectropolarimetric observations, they announced detections of a weak mean longitudinal magnetic field of the order of a few hundred Gauss in a dozen SPB stars and in the β Cephei star ξ^1 CMa (Hubrig et al. 2006a). The longitudinal magnetic field of the order of 400 G detected in ξ^1 CMa is currently the strongest among all known magnetic β Cephei stars.

The Stokes I and Stokes V spectra of ξ^1 CMa and the FORS 1 regression detection are presented in Figure 7.14.

Spectroscopic and photometric observations revealed that ξ^1 CMa pulsates monoperiodically and non-linearly in a radial mode with a period $P_{\rm puls} = 0.209\,574$ day and with a velocity amplitude of 33 km s^{-1} (Saesen et al. 2006). Morel et al. (2006) performed an abundance analysis of nine β Cephei stars and discovered that ξ^1 CMa is nitrogen enriched, as well as three other β Cephei stars: δ Cet, V2052 Oph, and β Cep. Notably, two of them, V2052 Oph and β Cep, have a detectable longitudinal magnetic field of the order of ~100 G.

A second systematic survey of massive pulsating stars was carried out in the years 2006–2008 and was based on 98 FORS 1 spectropolarimetric observations of 13 β Cephei stars, 34 SPB stars and seven normal B stars (Hubrig et al. 2009a). The results of these two surveys for the first time demonstrated that β Cephei and SPB stars can no longer be considered as classes of non-magnetic pulsators. Out of the 13 β Cephei stars studied with FORS 1, four showed weak magnetic fields. Among them, a magnetic field was detected in the β Cephei SB2 system ϵ Lup, which recently received much attention due to the presence of a magnetic field in both components. The fraction of magnetic SPBs and magnetic candidate SPBs was found to be higher: roughly half of the 34 SPB stars were found to be magnetic. The strength of the detected mean longitudinal magnetic fields was mainly in the range from 200 down to below 100 G (Hubrig et al. 2009a).

Many pulsating stars show strong spectral variability, which has an impact on the magnetic field measurements. As an example, a FEROS time series of the β Cephei star V1449 Aql showed line profile variability with a pulsating frequency of $f_{\rm puls} = 5.487$ day^{-1}. The variability in the spectra of V1449 Aql and the impact of the pulsations on the polarimetric spectra are presented in Figure 7.15. Neglecting strong variability of line profiles on short timescales can lead to a non-detection of magnetic fields in these stars, if the whole sequence of sub-exposures at different retarder waveplate angles has a duration comparable to the timescale of the pulsation variability (Hubrig et al. 2011b).

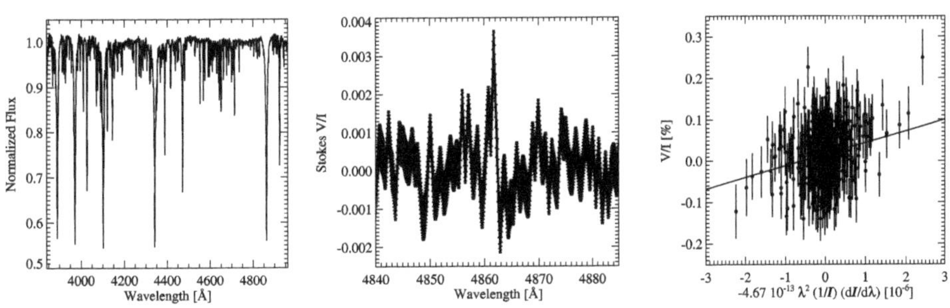

Figure 7.14. Stokes I spectrum (*left panel*), Stokes V spectrum (*centre panel*) and regression detection (*right panel*) for ξ^1 CMa. The thickness of the plotted line in the Stokes V spectrum corresponds to the uncertainty of the measurement of polarisation, determined from photon noise. Credit: Figure 1 from Hubrig et al. (2006a).

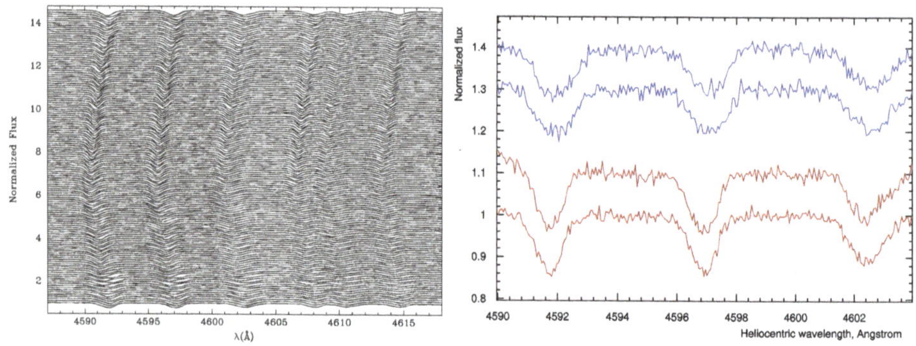

Figure 7.15. *Left panel:* time series of FEROS spectra for V1449 Aql showing pulsational line profile variability in the spectral region 4590–4615 Å. The pulsation phase zero is at the bottom. *Right panel:* variability of the output spectra in two SOFIN sub-exposures taken with the quarter-waveplate angles separated by 90° taken around HJD 2 455 398.530. The lower two spectra, $(I + V)_0$ and $(I - V)_0$, correspond to the first sub-exposure, while the upper spectra, $(I - V)_{90}$ and $(I + V)_{90}$, correspond to the second sub-exposure. The strong effect of pulsations on the line profile shapes and the line positions is clearly visible between the spectra of the first sub-exposure with a duration of 20 min and the spectra of the second sub-exposure with the same duration. Credit: Hubrig et al. (2011b), reproduced with permission © ESO.

Strong variability of Stokes I and Stokes V spectra in ESPaDOnS polarimetric spectra of the β Cephei star ξ^1 CMa is presented in Figure 7.16. These observations were acquired from 2017 to 2019 and show the change of the field polarity from positive to negative in observations carried out in 2018. The Stokes I and Stokes V profiles were calculated using the LSD technique described in detail by Donati et al. (1997). The line sample used consisted of 105 photospheric absorption lines retrieved from the Vienna Atomic Line Database (VALD; e.g., Kupka et al. 2011), adopting the stellar parameters $T_{\rm eff} = 27\,000$ K and $\log g = 4.0$ from the work of Shultz et al. (2017). The strong impact of pulsations on the Stokes I and Stokes V profiles is clearly visible. In particular, the LSD Stokes V profiles almost completely disappear at several observing epochs, leading to non-detections of the longitudinal magnetic field.

To take into account the impact of pulsations on the magnetic field measurement, a specific measurement strategy was suggested by Hubrig et al. (2011b). Before combining all sub-exposures obtained in high-resolution spectropolarimetric observations, each individual spectrum for each sub-exposure obtained at a different retarder waveplate angle has to be corrected by the measured radial velocity shift caused by pulsations. However, as presented in Figure 7.15, β Cephei pulsational line profile variability, in addition to radial velocity shifts, also has an impact on the line profile shape. Therefore, it is advisable to calculate the longitudinal magnetic field value as an average of the measurements for each sub-exposure. Järvinen et al. (2017) showed that ignoring the impact of pulsations on the field measurements can lead to a significant underestimation of the longitudinal magnetic field strength. The observations acquired for ξ^1 CMa in 2018 and 2019 presented in Figure 7.16 convincingly show that an impact of pulsations can even lead to a nearly complete cancellation of the magnetic signal in the Stokes V profiles.

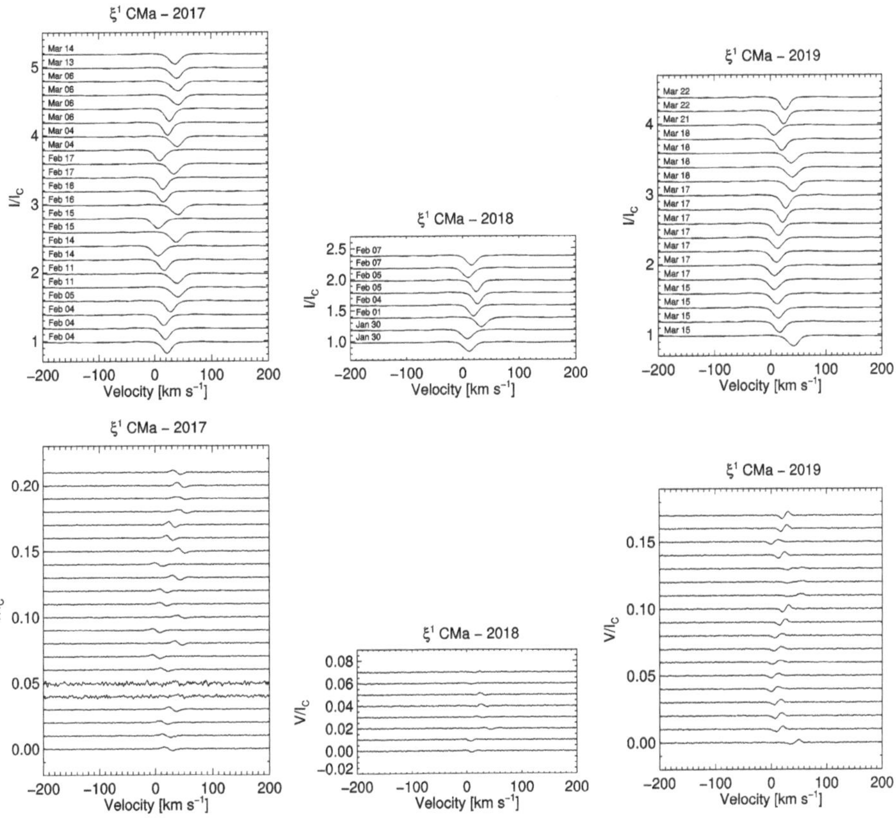

Figure 7.16. *Upper panel:* LSD Stokes *I* profiles of ξ^1 CMa calculated using high-resolution ESPaDOnS spectropolarimetric observations obtained from 2017 to 2019. *Lower panel:* corresponding LSD Stokes *V* profiles showing strong variability related to β Cephei pulsations causing at some epochs an almost complete disappearance of the magnetic signal.

A number of β Cephei stars exhibit weak emission lines probably caused by non-LTE effects or by optical pumping by far-UV lines (Rivinius et al. 2015). Studies of pulsational variability of Zeeman features in Stokes *V* profiles of individual lines formed in different atmospheric depths is a powerful tool to study the dynamical structure of magnetised atmospheres (Nardetto et al. 2013). The Stokes *I* and Stokes *V* profiles of the emission line Si III λ8103 recorded in the ESPaDOnS spectra of ξ^1 CMa acquired in 2019 are presented in Figure 7.17. However, Zeeman features are hardly detectable in the Stokes *V* profiles, appearing mainly at the level of noise. Obviously, higher quality and longer time series are needed to utilise the usefulness of emission lines in massive pulsating stars.

Quite unexpectedly, observations of ξ^1 CMa with the XMM-Newton telescope revealed the presence of X-ray pulsations with the same period as the stellar radial pulsation (Oskinova et al. 2014). The observed X-ray light curve and the X-ray/optical light curves phased with the pulsation period are presented in Figures 7.18 and 7.19. This first discovery of X-ray pulsations from a non-degenerate, massive star stimulates

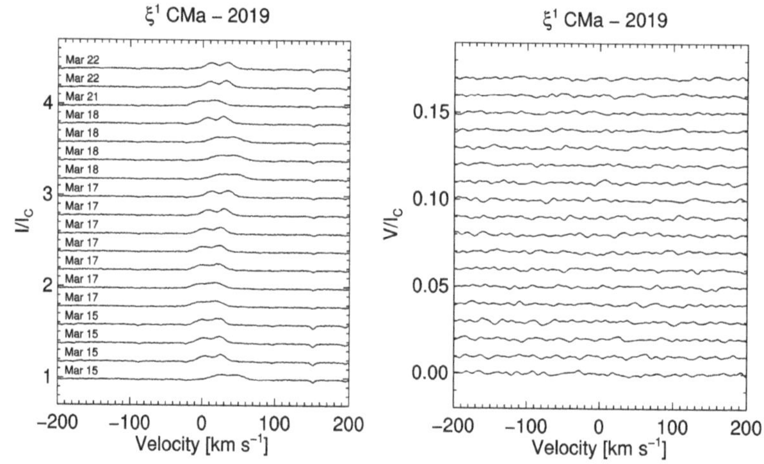

Figure 7.17. Stokes I (*left panel*) and Stokes V (*right panel*) spectra of ξ^1 CMa in the vicinity of Si III λ8103 obtained for ESPaDOnS observations in 2019.

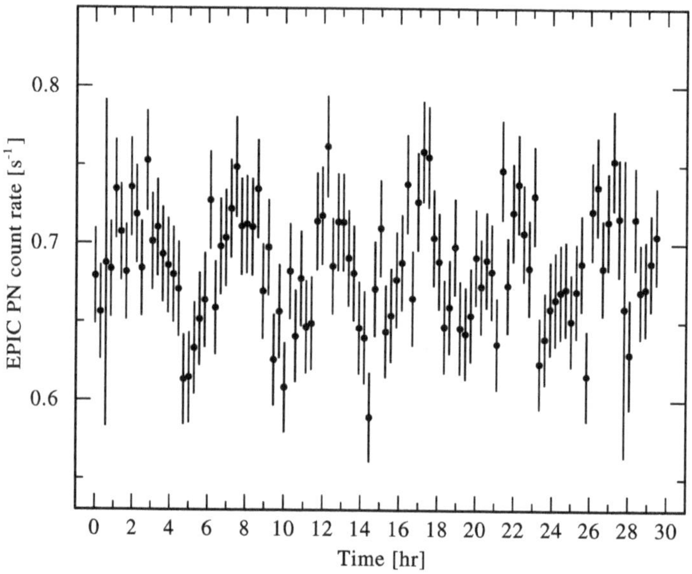

Figure 7.18. X-ray light curve of ξ^1 CMa in the 0.2–10.0 keV (1.24–62 Å) energy band binned to 1000 s with the background subtracted. The horizontal axis denotes the time after the beginning of the observation in hours. The vertical axis shows the count rate as measured by the EPIC PN camera. The 1σ error bars correspond to the combination of the error in the source counts and the background counts. Credit: Figure 2 from Oskinova et al. (2014).

theoretical considerations for the physical processes operating in magnetised stellar winds.

In an attempt to understand why only a fraction of the pulsating stars exhibit magnetic fields, Hubrig et al. (2009a) studied the position of magnetic and non-magnetic pulsating stars in the H-R diagram. The authors found that pulsating and

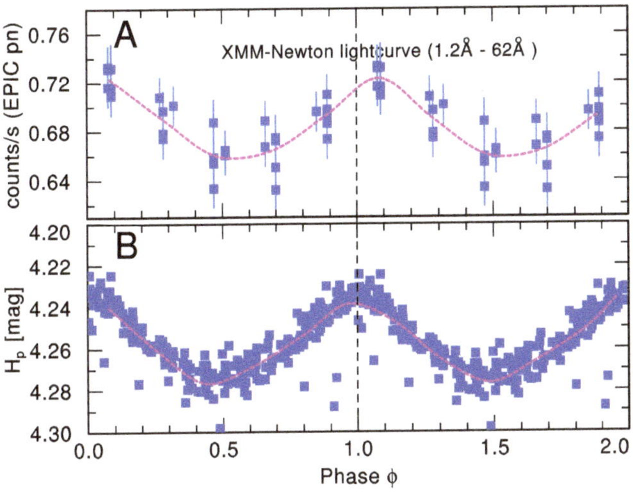

Figure 7.19. X-ray (*upper panel*) and optical (*lower panel*) light curves of ξ^1 CMa, phased with the stellar pulsation period. The X-ray light curve with 1 h binning is produced using the data obtained with the XMM-Newton EPIC PN camera. The dashed red line in the upper panel represents an interpolation of the averages in phase bins of $\Delta\phi = 0.1$. The lower panel shows the Hipparcos Catalogue Epoch Photometry data. The abscissa is the magnitude H_p in the Hipparcos photometric system (330–900 nm with the maximum at ~420 nm). The dashed red line represents an interpolation of the averages. Credit: Part of Figure 3 from Oskinova et al. (2014).

magnetic stellar domains in the H-R diagram largely overlap and no clear picture emerges as to the possible evolution of the magnetic field across the main sequence. The analysis of a representative sample of pulsating stars indicated that it is possible that stronger fields tend to be found in stars with lower pulsating frequencies and smaller pulsating amplitudes. Another trend was found if a correlation between the field strength and the v sin i-values was considered, i.e., stronger magnetic fields tend to be detected in more slowly rotating stars. Hints of some loose relations between the magnetic field strength, the dominant frequency and the corresponding amplitude were found, though additional magnetic studies are still needed to confirm these trends.

Although magnetic fields have been detected in a number of β Cephei and SPB stars, there are only very few studies on seismic modelling of magnetic pulsating stars. Forward seismic modelling of the slowly rotating β Cephei B2IV/V star V2052 Oph (= HD 163472) was presented by Briquet et al. (2012). A large-scale magnetic field is usually thought to suppress the near-core mixing caused by convective core overshooting in massive stars. The analysed data set of V2052 Oph included more than 1300 spectra from 10 different telescopes around the world and the resultant frequency spectrum included a radial mode and two prograde non-radial modes identified by means of spectroscopy. Forward seismic modelling of these identified pulsation modes yielded a relatively fast rotation rate and only a small amount of convective core overshooting, which was concluded to be low because of the star's magnetic field (Briquet et al. 2012).

In a more recent work, Buysschaert et al. (2017) presented forward seismic modelling of the pulsating magnetic B3.5V star HD 43317. This star exhibits g and p modes based on its frequency spectrum. Thus it appeared to have independent pulsation modes in both high and low frequency regimes. The modelling of Buysschaert et al. (2017) resulted in a rather low value for the convective core overshooting parameter, compatible with the suppression of near-core mixing due to a magnetic field. However, the uncertainties were too large to pinpoint such suppression as the sole physical interpretation.

The major shortcoming of current stellar evolution models for massive stars is that they contain theoretical uncertainties. The mixing profile at the interface of convective and radiative regions and the mixing profile within the envelope directly impact the amount of hydrogen available for nuclear burning. With more internal mixing, a massive star experiences a longer main-sequence life and produces a larger helium core mass at the end of the main sequence (e.g., Michielsen et al. 2019). Therefore, a numerical implementation of convection and convective-boundary mixing in magnetic and non-magnetic stars is crucial in determining core masses and the subsequent evolution of these stars.

7.6 Be Stars

The majority of emission line B-type stars constitute the so-called Be stars, which are defined as rapidly rotating main sequence stars showing normal B-type spectra with superimposed Balmer line emission. These stars lose mass and initially accumulate it in a rotating circumstellar disk. In addition, Be stars are characterised by the episodic dissipation and formation of a new circumstellar (CS) disk-like environment, non-radial pulsations and photometric and spectroscopic variability. Much of the mass loss is in the form of outbursts and thus additional mechanisms such as the beating of non-radial pulsation modes or magnetic flares must be at work.

A number of physical processes in classical Be stars (e.g., angular momentum transfer to a CS disk, channelling of stellar wind matter, accumulation of material in an equatorial disk, etc.) are more easily explainable if magnetic fields are invoked (e.g., Brown et al. 2004). Cassinelli et al. (2002) suggested a magnetically torqued disk model, in which a sufficiently strong magnetic field of the order of a few hundred Gauss channels a flow of wind material toward the equatorial plane to form a disk. Maheswaran (2003) developed the magnetic rotator wind disk model, in which Keplerian disks may be formed by magnetic fields of the order of a few tens of Gauss. Maheswaran & Cassinelli (2009) obtained solutions for the structure and the evolution of a protodisk region, i.e., the disk region that is initially formed when wind material is channelled by a dipole-type magnetic field toward the equatorial plane, showing that magneto-rotational instability may assist in the formation of a quasi-steady disk. According to their calculations, magnetic fields of the order of a few tens of Gauss will be able to channel a wind flow into a protodisk region.

The presence of magnetic fields in Be stars is highly controversial and still under debate. Indirect evidence for the presence of a magnetic field are variations of X-ray emission and transient features in absorption line profiles. Also angular momentum transfer to a circumstellar disk, channelling stellar wind matter and the accumulation of material in an equatorial disk are more easily explained if magnetic fields are brought into account. Due to the high rotation of Be stars and the presence of strong Balmer emission lines, magnetic field measurements are difficult and rare.

Fifteen Be stars were measured with a hydrogen polarimeter by Barker et al. (1985) using the Hβ line, but no detection was achieved. The earlier reported magnetic field detection in ω Ori employing high-resolution spectropolarimetry by Neiner et al. (2003c), was not confirmed by subsequent observations. Weak longitudinal magnetic fields at a significance level $>3\sigma$ were diagnosed in low-resolution spectropolarimetric FORS 1 observations of the three Be stars HD 56014 (= EW CMa; Hubrig et al. 2007), HD 148184 (= χ Oph; Hubrig et al. 2008c), and HD 208057 (= 16 Peg; Hubrig et al. 2006a). HD 208057 has a v sin i = 104 km s^{-1} and was classified as a Be star by Merrill & Burwell (1943) due to the detection of double emission in Hα. Later on, Henrichs et al. (2009) confirmed the presence of a magnetic field in HD 208057 using measurements acquired with the spectropolarimeters Narval and ESPaDOnS in 2007. However, the presence of emission in Hα was not detected in these observations. Thus, the question whether HD 208057 is indeed a classical Be star remains open.

Additional observations of Be stars using low-resolution FORS 1 spectropolarimetry were acquired from 2006 to 2008. Weak photospheric magnetic fields were reported in four field Be stars: HD 62367, μ Cen, o Aqr, and ϵ Tuc (Hubrig et al. 2009b). Stokes I and Stokes V spectra in the region including the Hδ and Hγ lines of the Be star o Aqr with $\langle B_z \rangle = 98 \pm 31$ G are displayed in Figure 7.20. The detected magnetic fields are usually weak, below 100 G, with the strongest longitudinal magnetic field, $\langle B_z \rangle = 117 \pm 38$ G, detected using hydrogen lines in the Be star HD 62367.

The possible presence of weak magnetic fields below 100 G was also discussed by Grunhut & Neiner (2015), who reported on Narval and ESPaDOnS high-resolution spectropolarimetric observations of about 80 Be stars in the framework of the MiMeS survey. The observations of this sample resulted in no direct evidence of a magnetic field. However, a rather low magnetic sensitivity for the majority of these observations with a median uncertainty of 103 G was insufficient to detect weak magnetic fields below 100 G. Importantly, high-resolution spectropolarimetric observations of very fast rotating Be stars acquired on telescopes with rather small apertures, such as the 2 m TBL Telescope of the Pic du Midi and the 3.6 m CFHT telescope fail to show credible measurement results if the whole sequence of sub-exposures at different retarder waveplate angles including overheads has a duration comparable to the timescale of the rotational variability or to the length of pulsation periods. Since rotation periods can be as short as a couple of hours (e.g., Ozuyar et al. 2018), high-resolution observations lead to phase smearing and a much lower magnetic signal. Obviously, observations using low-resolution spectropolarimeters operating at large telescopes are best suited for studies of the presence of magnetic fields in fast rotating Be stars exhibiting in addition non-radial pulsations.

Several Be stars exhibit X-ray flares and a rapid variability of absorption features called dimples (Smith et al. 2016, 1993). Dimples are believed to be produced from

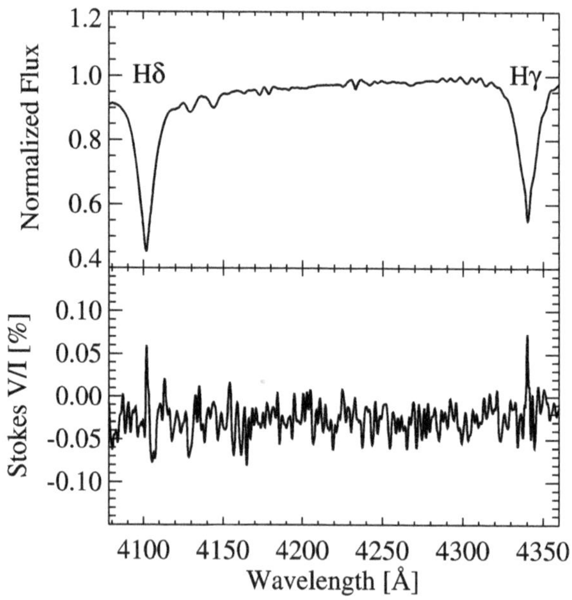

Figure 7.20. Stokes *I* (*top*) and Stokes *V* (*bottom*) spectra of the Be star *o* Aqr with $\langle B_z \rangle = 98 \pm 31$ G in the region including the Hδ and Hγ lines. Credit: Figure 5 from Hubrig et al. (2009b).

an ablation of photospheric material caused by a nearby flare (e.g., Smith et al. 1993). Since the magnetic field topology on the surface of fast rotating Be stars is unknown, it is reasonable to study the polarisation signal induced in the spectral lines by the Zeeman effect through long time series of exposures with short integration times. This also permits to search for the presence of local transient magnetic fields on the surface of these stars. The magnetic rotator wind disk model presented by Maheswaran (2003) does not require large-scale organised magnetic fields, axially symmetric fields or uniformly strong fields across the entire stellar surface. It qualitatively applies to stars with magnetic fields consisting of flux loops that emerge from lower latitudes and thread the disk around the Be star. On the other hand, the detection limit of local fields depends critically on the structure of the magnetic regions, which is not known a priori.

Time-resolved magnetic field measurements carried out over about one hour (with up to 30 magnetic field measurements) using FORS 1 at the VLT were acquired for nine early-type Be stars by Hubrig et al. (2009b). Using magnetic field measurements obtained for the star λ Eri, a periodicity with $P = 21.1$ min was detected. The resulting phase diagram and amplitude spectrum are presented in Figure 7.21. The periodicity of the time-resolved magnetic field measurements was analysed by the application of Breger's code (Breger 1990). Using FORS 1 spectropolarimetric observations acquired in 2006 August, a 2.4σ peak corresponding to a period of 21.12 min was detected in the data set of the magnetic field measurements carried out using the hydrogen lines. This peak appears with a 2.2σ in the data set of measurements carried out using the entire spectra. Notably, spectral line profiles of λ Eri exhibit short-time periodic variability due to non-radial

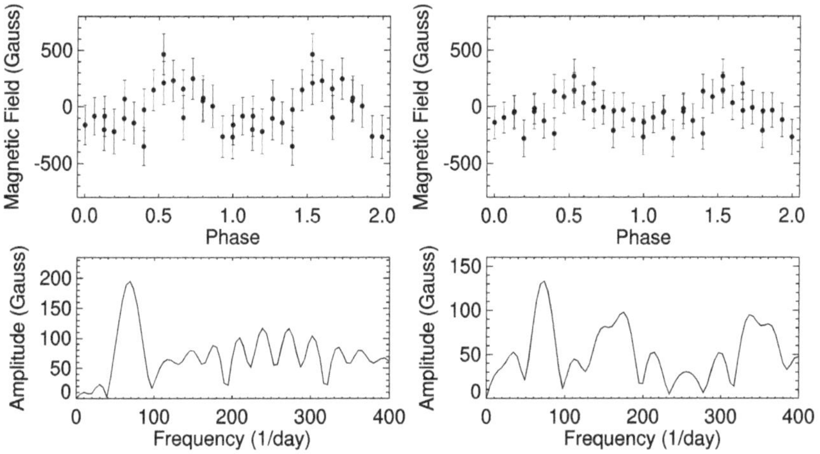

Figure 7.21. Periodicity detected in time-resolved FORS 1 spectropolarimetric observations of λ Eri acquired in 2006 August. Phase diagram and amplitude spectrum for the magnetic field measurements using hydrogen lines (*left panel*) and all lines (*right panel*). Credit: Figure 1 from Hubrig et al. (2009b).

pulsations with a period of 0.7 day (Kambe et al. 1993). A typical variability pattern in a few line profiles in FORS 1 spectra recorded at three different epochs in 2006 and 2007 is presented in Figure 7.22. Apart from pulsational variability, also dimples with a duration of 2–4h were detected by Smith (1994). An additional spectropolarimetric time series of λ Eri with FORS 2 over three hours was obtained in 2016 (Hubrig et al. 2017). A search for periodicity in the combined time series from 2006 to 2016 indicated the potential presence of a magnetic variability modulation with a period of 13.7 min, which possibly can be taken as a hint for the presence of a small-scale magnetic field on the stellar surface. Apart from λ Eri, four other stars showed indications of magnetic cyclic variability on the scales of tens of minutes (Hubrig et al. 2009b). Obviously, massive Be stars are good candidates for future time-resolved magnetic field observations with more efficient spectropolarimeters.

It is of interest that a short-term magnetic field periodicity with $P = 8.8$ min was also detected for the B0 star θ Car (Hubrig et al. 2008a), which shows abundance peculiarities such as enhanced N and depleted C and is considered to be a blue straggler, one of a few known in young open clusters. In this scenario, this star would be a post-Roche lobe overflow system.

According to the work of Ozuyar et al. (2018), the light variations of the Be star 48 Lib can be interpreted as a rotational modulation caused by an inhomogeneous region related to a co-rotating spot region. The existence of surface spots was theoretically discussed by Cantiello & Braithwaite (2011) and observationally by Balona (2016). Cantiello & Braithwaite (2011) suggested that dynamo action in thin subsurface convection zones in hot stars might generate a small-scale magnetic field. In such a case, magnetic spots can manifest themselves as hot, bright spots.

Hubrig et al. (2009b) used the FORS 1 spectropolarimetric mode in 2006–2008 to observe a sample of Be stars in the open young cluster NGC 3766 in the Carina

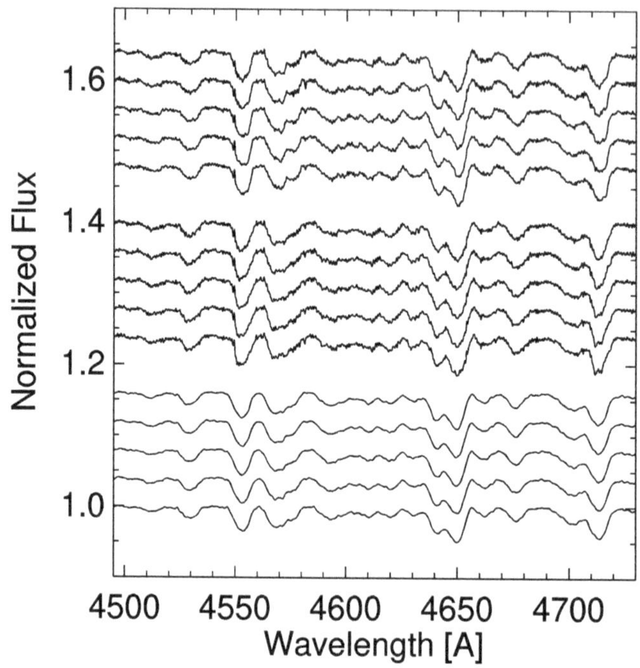

Figure 7.22. Spectrum variability of λ Eri on three different nights: on 2006 August 8 (*bottom*), 2007 November 27 (*middle*), and 2007 November 28 (*top*). Credit: Figure 4 from Hubrig et al. (2009b).

spiral arm. This cluster turned out to be extremely interesting, with evidence for the presence of a magnetic field in seven early B-type stars out of the observed fourteen cluster members. Three stars with detected magnetic fields were classified in the literature as Be stars or Be star candidates. Examples of the observed Stokes I and Stokes V profiles of a He-rich member of this cluster with a measured longitudinal magnetic field $\langle B_z \rangle = 1559 \pm 38$ G and of a member classified as a potential Be star by Shobbrook (1985) with a measured longitudinal magnetic field $\langle B_z \rangle = -194 \pm 62$ G are presented in Figure 7.23. The magnetic fields were detected in stars with $T_{\rm eff}$ between 15 500 K and 21 435 K and $\log g$ from 4.61 to 3.51, indicating that the presence of a magnetic field is not directly related to the stellar evolutionary phase on the main sequence.

To summarise, magnetic fields in rapidly rotating Be stars, characterised by episodic dissipation and formation of a CS disk-like environment, non-radial pulsations and photometric and spectroscopic variability, are especially hard to detect because these fields, if indeed present, are expected to be weak, of the order of less than 100 G (e.g., Hubrig et al. 2009b). For the first time, several time series with the low dispersion spectrographs FORS 1/2 in spectropolarimetric mode mounted at one of the biggest telescopes in the world were combined to investigate the variability of the longitudinal magnetic field on the surface of a Be star. Certainly, the potential detectability of complex magnetic fields, characterised by random spot distributions on the surface of Be stars, needs to be further investigated

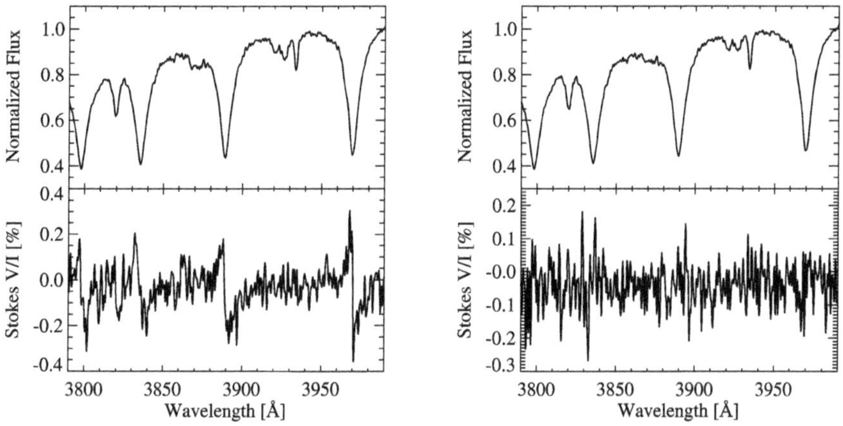

Figure 7.23. *Left panel:* FORS 1 Stokes I and Stokes V spectra obtained for the He-rich star NGC 3766-170 with a magnetic field $\langle B_z \rangle = 1559 \pm 38$ G. *Right panel:* Stokes I and Stokes V spectra obtained for the candidate Be star NGC 3766-45, with a magnetic field $\langle B_z \rangle = -194 \pm 62$ G. Credit: Figure 8 from Hubrig et al. (2009b).

by excessive spectropolarimetric time series. Since the majority of Be stars exhibit non-radial pulsations, which cause significant changes in line profiles and their positions on rather short timescales, such time series should be predominantly carried out with low-resolution spectropolarimeters, where only very short exposure times at different retarder waveplate angles are required.

7.7 High-mass X-ray Binaries

Among the bright X-ray sources in the sky, a significant number consist of a compact object (either a neutron star or a black hole) accreting from the wind of a star with mass above 10 $M_{\odot}$. Such systems are called high-mass X-ray binaries (HMXBs). They are young (several dozen million years old) and can be formed when one of the initial members loses a significant part of its mass through stellar wind or mass transfer, before the first supernova explosion occurs (van den Heuvel & Heise 1972). Studies of different types of HMXBs are of special interest to obtain reliable predictions about the populations of relativistic binaries. Furthermore, high-mass X-ray binaries are fundamental for studying stellar evolution, nucleosynthesis, structure, and evolution of galaxies and accretion processes.

7.7.1 Classical HMXBs with Supergiant Companions

The classical HMXBs with supergiant companions (sgHMXBs) are known since the birth of X-ray astronomy. They are persistent X-ray emitters, with a limited range of intensity variability around a factor of 10. These targets are among the brightest X-ray sources in the sky. For these sources, the observed spectral and time variability is best explained by assuming that accretion onto the compact object is taking place from a highly structured stellar wind, where cool dense clumps are embedded in a rarefied photoionized gas.

Previous circular polarisation observations obtained with FORS 1 at the VLT in 2007/2008 revealed the presence of a weak longitudinal magnetic field on the surface of the optical component of the sgHMXB Cyg X-1, which contains a black hole and an O9.7Iab supergiant on a 5.6 days orbit (Karitskaya et al. 2010). In a sample of five black hole HMXB systems known in the Galaxy, Cyg X-1 is the best studied system due to its brightness. Karitskaya et al. (2010) detected a relatively weak mean longitudinal magnetic field of the order of 100 G with a few measurements at a significance level in the range between 3.5 and 6.2σ.

The magnetic field of the eclipsing sgHMXB Vela X-1 with a B0.5Ia component and an orbital period of 8.96 days was studied by Hubrig et al. (2013). This system can be considered as the prototype of persistent HMXBs. No magnetic field detection at a significance level of 3σ has been achieved in this system. The spectral behaviour of Vela X-1 is known to be very complex due to the presence of bumps and wiggles in the line profiles and the impact of tidal effects producing orbital phase-dependent variations in the line profiles leading to asymmetries such as extended blue or red wings (Koenigsberger et al. 2012). Furthermore, Kreykenbohm et al. (2008) detected flaring activity and temporary quasi-periodic oscillations in INTEGRAL X-ray observations of this star. More recent spectropolarimetric observations were acquired with FORS 2 for Cyg X-1 and two other sgHMXB systems, BP Cru and LS 5039 (Hubrig et al. 2019a), indicating the presence of magnetic fields at a significance level of 2.3 to 2.5σ. The detection of a magnetic field at a significance level of 2.5σ in LS 5039 is particularly interesting, as it is known to be a source emitting non-thermal radiation. It is also a widely studied rather faint gamma-ray binary system with $m_V = 11.3$, hosting a main-sequence O-type star and probably a young non-accreting pulsar (e.g., Molina & Bosch-Ramon 2020). The spectral appearance of all targets studied with FORS 2 is presented in Figure 7.24.

Spectropolarimetric observations acquired with FORS 2 showed changes in the magnetic field polarities for all four sgHMXBs, but the measurement uncertainties were too large (Hubrig et al. 2019a). On the other hand, the presence of a magnetic field cannot be excluded for some of them: as shown in Figure 7.25, typical Zeeman features are detected in the FORS 2 Stokes V spectra of BP Cru and Cyg X-1. Apart from the magnetic field measurements, the acquired FORS 2 spectra of the optical components were used to study spectral variability. In the spectra of all targets, spectral lines belonging to hydrogen and other elements showed changes of the line intensities and radial velocities over different observing nights (Hubrig et al. 2019a). An example of the detected variability in helium and hydrogen line profiles in the spectra of BP Cru is presented in Figure 7.26. It is not clear yet whether the detected variability is caused by the presence of magnetospheres or by pulsational variability, frequently observed in OB-type stars. Further observations are needed to be able to draw solid conclusions about the role of magnetic fields in these targets.

7.7.2 Supergiant Fast X-ray Transients

A fraction of the high-mass X-ray binaries are supergiant fast X-ray transients (SFXTs). SFXTs are a subclass of the HMXBs, associated with early-type

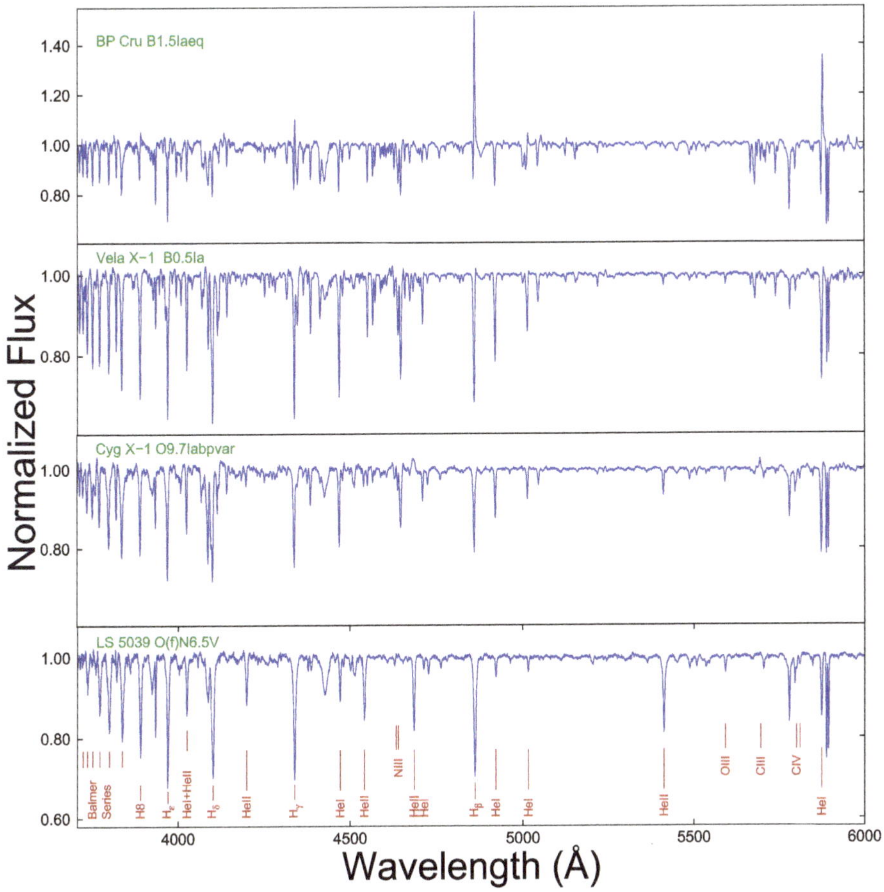

Figure 7.24. Normalised FORS 2 spectra of LS 5039, Cyg X-1, Vela X-1, and BP Cru (*from bottom to top*). Well known spectral lines are indicated. Credit: Figure 1 from Hubrig et al. (2019a).

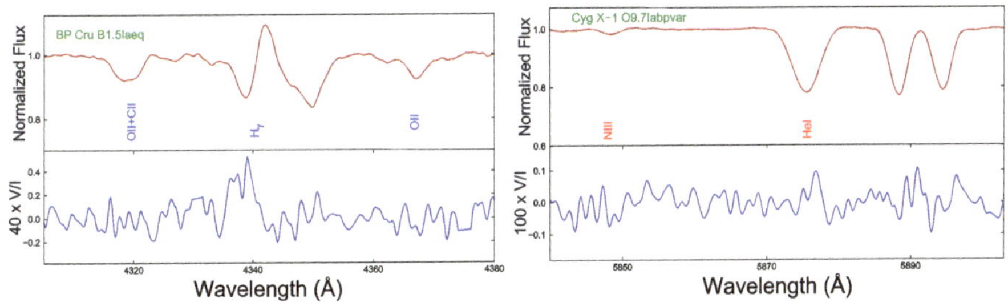

Figure 7.25. Examples of Stokes *I* and Stokes *V* spectra observed for BP Cru (*left panel*) and Cyg X-1 (*right panel*) in the vicinity of the Hγ line and the He I 5876 line, respectively. Credit: Figure 2 from Hubrig et al. (2019a).

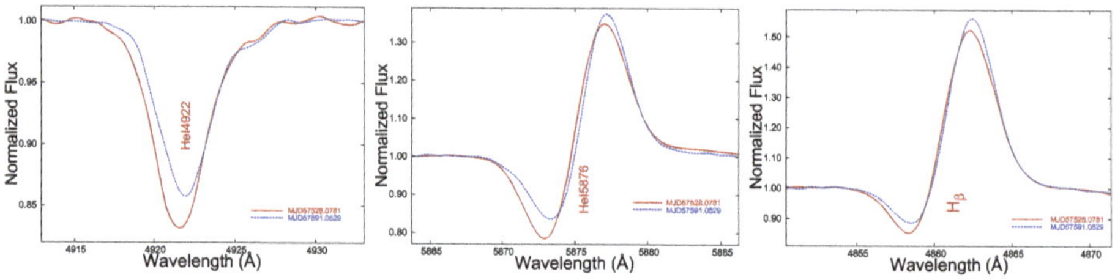

Figure 7.26. Variability of spectral lines in the FORS 2 spectra of BP Cru recorded on two different nights. Credit: Figure 3 from Hubrig et al. (2019a).

supergiant companions and characterised by sporadic, short and bright X-ray flares reaching peak luminosities of 10^{36}–10^{37} erg s^{-1} and typical energies released in bright flares of about 10^{38}–10^{40} erg (e.g., Sidoli 2017). Their X-ray spectra in outburst are very similar to accreting pulsars in HMXBs. In fact, half of them have measured neutron star spin periods similar to those observed from persistent HMXBs (Shakura et al. 2015; Martínez-Núñez et al. 2017).

The physical mechanism driving their transient behaviour, probably related to the accretion of matter from the supergiant wind by the compact object, has been discussed by several authors and is still a matter of debate. The leading model for the existence of SFXTs invokes their different wind properties and magnetic field strengths, which lead to distinctive accretion regimes (Shakura et al. 2012; Postnov et al. 2015; Shakura et al. 2015; Shakura & Postnov 2016). The SFXTs' behaviour can be explained by sporadic capture of magnetised stellar wind. The effect of the magnetic field carried by the stellar wind should be twofold: first, it may trigger rapid mass entry to the magnetosphere via magnetic reconnection in the magnetopause (a phenomenon that is well known on the dayside of the Earth's magnetosphere) and second, the magnetised parts of the wind (magnetised clumps with a tangent magnetic field) have a lower velocity than the non-magnetised parts (or the parts carrying the radial field; Shakura et al. 2015). The model predicts that a magnetised clump of stellar wind with a magnetic field strength of a few 10 G triggers sporadic reconnection, allows accretion and results in an X-ray flare. Typically, the neutron star orbital separation is a few R_*. Thus, the expected required magnetic field on the stellar surface of the supergiant is of the order of 100–1000 G.

To investigate the magnetic nature of these targets, Hubrig et al. (2018, 2019b) observed the two optically brightest SFXTs, IGR J08408−4503 ($P_{\rm orb}$ = 9.5 days) and IGR J11215−5952 ($P_{\rm orb}$ = 165 days), using FORS 2 in spectropolarimetric mode. The spectral appearance of IGR J08408−4503 and IGR J11215−5952 in the FORS 2 spectra is presented in Figure 7.27. No significant magnetic field was measured in the spectra of IGR J08408−4503, with the highest value $\langle B_z \rangle_{\rm hyd} = -184 \pm 97$ G at a significance level of 1.9σ. On the other hand, a definite magnetic field detection was achieved for IGR J11215−5952 in 2016 December with $\langle B_z \rangle_{\rm hyd} = 416 \pm 110$ G, with a significance at the 3.8σ level. The measurement obtained in 2016 May yielded $\langle B_z \rangle_{\rm hyd} = -978 \pm 308$ G at a significance level of 3.2σ (Hubrig et al. 2018). Stokes V spectra of IGR J11215−5952 obtained on these two nights in the spectral region around the Hβ line are presented in Figure 7.28, where the Stokes V Hβ profiles

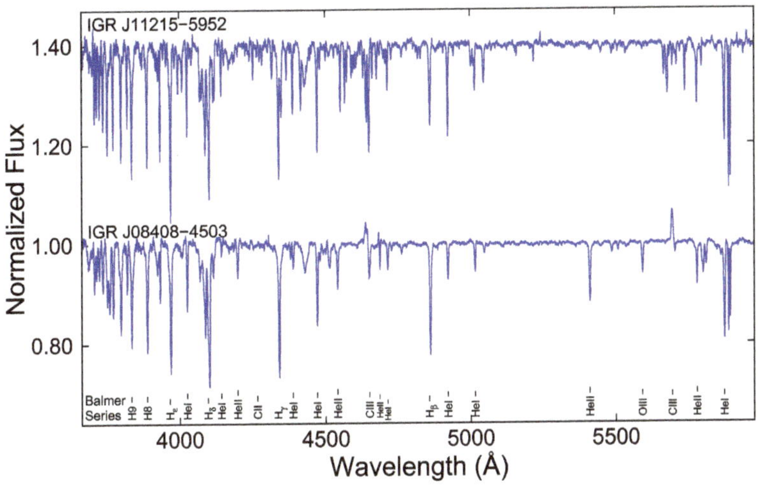

Figure 7.27. Normalised FORS 2 Stokes *I* spectra of IGR J08408−4503 and IGR J11215−5952. Well known spectral lines are indicated at the bottom. The spectrum of IGR J11215−5952 was vertically offset by 0.4 for better visibility. Credit: Figure 1 from Hubrig et al. (2018).

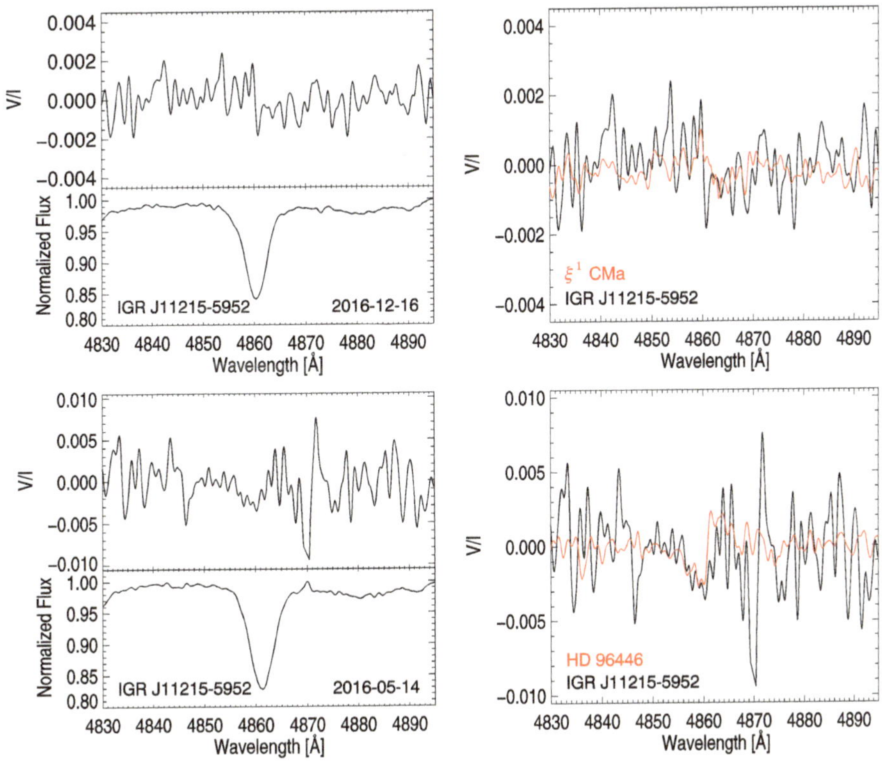

Figure 7.28. *Left panel*: Stokes *V* and Stokes *I* spectra of IGR J11215−5952 in the spectral region around the Hβ line at two different epochs. *Right panel*: Stokes *V* spectra of IGR J11215−5952 overplotted with the Stokes *V* spectra of the two well-known magnetic early B-type stars ξ^1 CMa ($\langle B_z\rangle_{\rm hyd}$ = 360 ± 49 G) and HD 96446 ($\langle B_z\rangle_{\rm hyd}$ = −1590 ± 74 G) for best visibility of the Zeeman features. Credit: Figure 2 from Hubrig et al. (2019b).

observed for IGR J11215−5952 are overplotted with the Stokes V Hβ profiles of the two well-known magnetic early B-type stars HD 96446 and ξ^1 CMa for the best visibility of the Zeeman features. Although no significant magnetic field was measured in the spectra of IGR J08408−4503, due to the presence of distinct Zeeman features in its spectra it appears still valuable to obtain additional spectropolarimetric observations on other epochs corresponding to different orbital phases.

IGR J11215−5952, for which spectropolarimetric observations revealed the presence of a magnetic field on two occasions, is the only SFXT where strictly periodic X-ray outbursts have been observed, repeating every 164.6 days (Sidoli et al. 2006, 2007; Romano et al. 2009). To explain these short periodic outbursts, Sidoli et al. (2007) proposed that they are triggered by the passage of the neutron star inside an equatorial enhancement of the outflowing supergiant wind, focused on a plane inclined with respect to the orbit. This configuration of a line-driven stellar wind might be magnetically channelled (ud-Doula & Owocki 2002). The effectiveness of the stellar magnetic field in focusing the wind is indicated by the wind magnetic confinement parameter η defined as $\eta = B_*^2 R_*^2 \, / \, (\dot{M}_{\rm W} \upsilon_\infty)$, where B_* is the strength of the magnetic field at the surface of the supergiant, R_* is the stellar radius ($R_* = 40\ R_\odot$), υ_∞ is the wind terminal velocity ($\upsilon_\infty = 1200$ km s^{-1}) and $\dot{M}_{\rm W}$ is the wind mass loss rate ($\dot{M}_{\rm W} = 10^{-6}\ M_\odot$ yr^{-1}; Lorenzo et al. 2014). Adopting $B_* \geqslant 0.7$ kG at the magnetic equator, the wind magnetic confinement parameter η becomes $\geqslant 500$, implying a wind confinement up to an Alfvén radius $R_A = \eta^{1/4} R_* \geqslant 4.73\ R_*$ (ud-Doula & Owocki 2002). This radial distance is compatible with the orbital separation at periastron in IGR J11215−5952, where the orbital eccentricity is high ($e > 0.8$; Lorenzo et al. 2014). The measured magnetic field strength in IGR J11215−5952 is therefore high enough to channel the stellar wind onto the magnetic equator, supporting the scenario proposed by Sidoli et al. (2007) to explain the short periodic outbursts in this SFXT.

In view of the probable detection of the magnetic field in IGR J11215−5952, it is necessary to enlarge the sample of SFXTs with studied magnetic fields. Presently, about 20 SFXTs are known in the Milky Way, but because of the faintness of the SFXTs—the majority has visual magnitudes $m_{\rm V} \geqslant 12$, up to $m_{\rm V} \geqslant 31$ (Sidoli 2017; Persi et al. 2015)—no high-resolution spectropolarimetric observations were carried out for these objects so far and the presented FORS 2 observations were the first to explore the magnetic nature of the optical counterparts. Future spectropolarimetric observations of a representative sample of SFXTs are urgently needed to be able to draw solid conclusions about the role of magnetic fields in the wind accretion process and to constrain the conditions that enable the presence of magnetic fields in massive binary systems.

7.8 Wolf–Rayet Stars

In the modelling of massive stars, specific aspects, especially the role of magnetic fields, remain not well understood, implying large uncertainties in the star's evolutionary path and ultimate fate. Previous observations indicate that probably about

7% of the O-type stars with masses exceeding 18 $M_{\odot}$ and about 6% of early B- and O-type stars have measurable, mostly dipolar magnetic fields (e.g., Grunhut et al. 2017; Schöller et al. 2017). Theoretical models suggest that O stars with strong magnetic field detections may be related to magnetars with $B \approx 10^{15}$ G (e.g., Thompson et al. 2004). Thus, it is important to detect magnetic fields in Wolf–Rayet (WR) stars, which are descendants of massive O stars and direct predecessors of compact remnants. More than 400 WR stars are known in the Galaxy. These stars are highly chemically evolved massive stars that have lost their hydrogen envelope and now expose their former stellar core. However, their dense winds make the stellar surface almost unobservable.

Magnetic fields in WR stars are currently not sufficiently explored, in spite of the fact that there is indirect evidence, e.g., from spectral variability and X-ray emission, that magnetic fields are present in WR star atmospheres (e.g., Michaux et al. 2014). Previous theoretical work by Gayley & Ignace (2010) predicted a fractional circular polarisation of a few times 10^{-4} for magnetic fields of about 100 G. Using high-resolution spectropolarimetric observations with ESPaDOnS at the Canada–France–Hawaii Telescope, de la Chevrotière et al. (2014) reported marginal magnetic field detections for WR 134, WR 137, and WR 138, corresponding to magnetic field strengths of about 200 G, 130 G, and 80 G, respectively, and an average upper limit of about 500 G for the non-detections in other stars. As the line spectrum in WR stars is formed in the strong stellar wind, the detection of magnetic fields in these stars is difficult. The major problem is the wind broadening of the emission lines by Doppler shifts with wind velocities of a few thousand km s^{-1}. The broad spectral lines observed with high-resolution spectropolarimetry extend over adjacent orders, so that it is necessary to adopt order shapes to get the best continuum normalisation.

Because of such immense line broadening in the spectra of WR stars, to search for weak magnetic fields in a number of WR stars, Hubrig et al. (2016) used the FOcal Reducer low dispersion Spectrograph (FORS 2) in spectropolarimetric mode on the VLT. Spectropolarimetric observations were acquired during two observing runs from 2010 to 2012. During the first run, Hubrig et al. (2016) observed the LMC WR star BAT99 7 and the Galactic stars WR 6, WR 7, WR 18, and WR 23. The second run was focused on monitoring WR 6.

A mean longitudinal magnetic field, $\langle B_z \rangle = 258 \pm 78$ G, at a significance level of 3.3σ was detected in the cyclically variable and X-ray emitting WN5 star WR 6. Keeping in mind that the two clearly magnetic Of?p stars HD 148937 and CPD −28 5104 were for the first time detected as magnetic in FORS 1/2 observations at significance levels of 3.1σ and 3.2σ, respectively (Hubrig et al. 2008c, 2011c, 2013), the detection of the magnetic field in WR 6 at a significance level of 3.3σ indicates that a magnetic field is very likely present in this star. Among the other targets, the highest value for the longitudinal magnetic field, $\langle B_z \rangle = 327 \pm 141$ G, was measured in the LMC star BAT99 7.

WR 6 is one of the most extensively observed WR stars, showing spectroscopic variability signatures that can be attributed to the presence of co-rotating interaction regions (CIR) in its wind. It also shows periodic optical and UV line and continuum variability with a period of $P_{rot} = 3.766$ days (e.g., Lamontagne et al. 1986). de la Chevrotière et al. (2013) reported magnetic measurements of WR 6 based on spectropolarimetric observations obtained with ESPaDOnS at CFHT. Although their measurements did not provide enough evidence to infer the presence of a magnetic field, a few observed Stokes V profiles showed organised patterns that fit well the predicted split monopole magnetic field model.

The spectropolarimetric monitoring of WR 6 with FORS 2 in 2011/2012 revealed a sinusoidal nature of the $\langle B_z \rangle$ variations, which is indicative of a predominantly dipolar magnetic field structure (Hubrig et al. 2016; see the left side of Figure 7.29). The field appeared to be reversing, with the extrema detected at rotation phases 0 and 0.5. In the work on WR 6, Hubrig et al. (2016) for the first time used another theoretical approach to estimate the magnetic field strength in stars with strong wind emission lines. Usually, in low-resolution circular polarisation observations of intermediate-mass stars, to estimate the strength of the longitudinal magnetic field, astronomers are searching for a correlation between V/I and $\mathrm{dln}I/\mathrm{d}\lambda$. For the specific situation with the spatially varying field in WR stars, such a correlation is only quantitatively defined, since the atmospheres in these stars are not static. As shown in Gayley & Ignace (2010), what is different for a magnetic field entrained in a hypersonic wind is that a correlation between the wind flow and the magnetic field direction induces a corresponding correlation between Doppler and Zeeman shifts, altering the correlation in ways that depend on the magnetic field model. To qualitatively understand the application of this correlation to wind lines in the

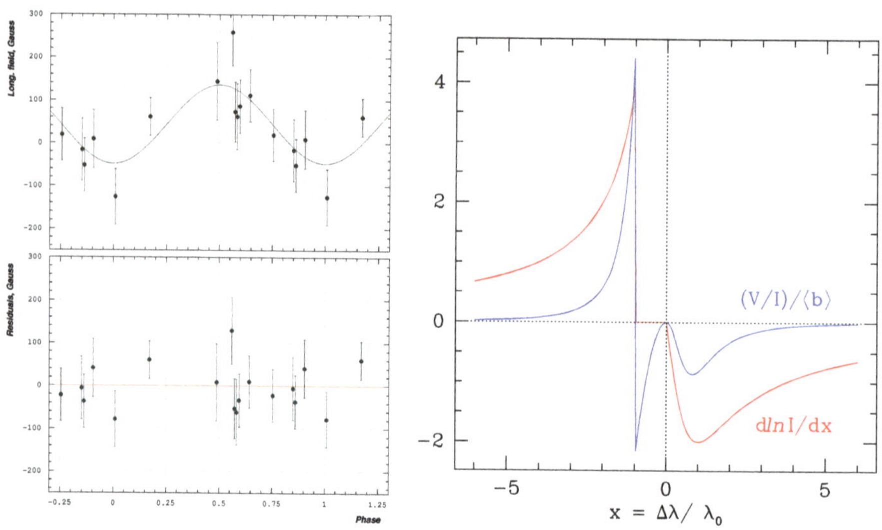

Figure 7.29. *Left panel*: phase diagram and best sinusoidal fit for the mean longitudinal magnetic field measurements of WR 6. The residuals (observed – calculated) are shown in the lower panel. *Right panel*: using the expression from Gayley & Ignace (2012) for a recombination wind emission line threaded by a split monopole magnetic field, $\mathrm{dln}I/\mathrm{d}x$ (red line) and $(V/I)/\langle b \rangle$ (blue line) are plotted against the normalised wavelength x. The case shown is for an optically thick line with $p = 6$. Credit: Figure 3 and parts of Figure 5 from Hubrig et al. (2016).

atmospheres of WR stars, the schematic quality of this correlation in winds is presented in Figure 7.29 on the right side, following the model presented in Gayley & Ignace (2012). In this figure, x is the line width in units of the radial Doppler shift at the wind photosphere and p sets the power law for the radial falloff of the volume emissivity. This approach was recently elaborated in more detail by Gayley (2017).

In the sample of spectropolarimetrically studied WR stars by Hubrig et al. (2016), WR 6 was the only target showing X-ray emission and cyclical variability due to the presence of CIRs, which are formed out of the interaction between high and low-velocity flows as the star rotates (e.g., St-Louis et al. 1995). CIRs were detected in spectroscopic time series observations of only a few massive stars (e.g., Mullan 1984). It was suggested that CIRs are related to the presence of magnetic bright spots, which are indicators of the presence of a global magnetic field (e.g., Ramiaramanantsoa et al. 2014). To substantiate expectations on the presence of magnetic fields in the most-evolved massive stars, Hubrig et al. (2020b) searched for magnetic fields in two other promising targets, WR 46 (= HD 104994) and WR 55 (= HD 117688), which, similar to WR 6, show CIRs (e.g., Chené & St-Louis 2011). These stars are accessible from the VLT on Cerro Paranal and have never been previously observed using spectropolarimetry.

WR 46 is a WN3p star (Hamann et al. 2019), with relatively strong O IV λ3811 and λ3834 emission lines. It is very hot and compact ($T_{\rm eff}$ = 112.2 kK, R_* = 1.4 $R_\odot$) and also bright in X-rays, possessing a hard component in its X-ray emission (Gosset et al. 2011), which may indicate the magnetic nature of WR 46. According to Hénault-Brunet et al. (2011), WR 46 is known to exhibit a very complex variability pattern. The different periods and timescales observed in the past suggested the presence of multiple periods, including dominant and secondary periods. To explain the short-term variability of this star, various scenarios were evoked, including the possibility of a close binary or non-radial pulsations (e.g., Veen et al. 2002a, 2002b). Using observations with the Far Ultraviolet Spectroscopic Explorer (FUSE), Hénault-Brunet et al. (2011) found significant variations on a timescale of ~8 h. This period is close to the photometric and spectroscopic periods previously reported by other authors. Hénault-Brunet et al. (2011) also reported the detection of a second significant peak, just slightly weaker, corresponding to $P = 15.5 \pm 2.5$ h.

WR 55 is a significantly cooler ($T_{\rm eff}$ = 56.2 kK) WN7 star with hydrogen deficiency and belongs to the WNE subclass, like WR 6 (Hamann et al. 2006). However, its radius, R_* = 5.2 $R_\odot$, is larger compared to the radius of WR 6 with R = 3.2 $R_\odot$. A highly significant level of spectroscopic variability of about 10% was discovered by Chené & St-Louis (2011). So far, no periodicity search was carried out for this star. Cappa et al. (2009) investigated the distribution of molecular gas related to the ring nebula RCW 78 around WR 55. They concluded that WR 55 is not only responsible for the ionisation of the gas in the nebula, but also for the creation of the interstellar bubble. Their analysis indicates that the star formation activity in the environment of this nebula may have been triggered by the expansion of the bubble. Thus, the discovery of a magnetic field in WR 55 would be of exceptional interest for star formation theories.

The spectral appearance of the WN stars WR 46 and WR 55 in the FORS 2 spectra is presented in Figure 7.30. The spectra of WN stars are dominated by helium and nitrogen lines. The WN3p star WR 46 is characterised by the presence of strong N v and He ii lines and the absence of hydrogen. The "p" stands for peculiar and denotes the presence of unusually strong O vi $\lambda 3811$ and $\lambda 3834$ emission lines, the relatively strong N v $\lambda 4604$ line and relatively weak C iv $\lambda 5801$ and $\lambda 5812$ lines (e.g., Conti & Massey 1989).

No field detection at a significance level of 3σ was achieved for WR 46, but the values for the longitudinal magnetic field $\langle B_z \rangle$ showed change of polarity, with the strongest mean longitudinal magnetic field of positive polarity $\langle B_z \rangle = 342 \pm 154$ G at a significance level of 2.2σ and the strongest field of negative polarity $\langle B_z \rangle = -199 \pm 88$ G at a significance level of 2.3σ. Since significant photometric and spectroscopic variations on a timescale of ~8 h were reported in previous studies of this star, assuming that this periodicity is caused by rotational modulation, Hubrig et al. (2020b) tested the distribution of the measurement values over this period. No hint for sinusoidal modulation, which is expected for a large-scale organised dipole field structure was detected. However, a rotational modulation was indicated in the magnetic field values with the period of 15.5 h suggested by Hénault-Brunet et al. (2011). Obviously, the fidelity of the 15.5 h period needs to be confirmed with a long-term extensive data set.

For WR 55, the values for the longitudinal magnetic field $\langle B_z \rangle$ showed a change of polarity, with the highest field values $\langle B_z \rangle = -378 \pm 85$ G at a significance level of 4.4σ and $\langle B_z \rangle = 205 \pm 58$ G at a significance level of 3.5σ.

The simplest model for the magnetic field geometry in stars with globally organised fields is based on the assumption that the studied stars are oblique dipole rotators, i.e., their magnetic field can be approximated by a dipole with the magnetic axis inclined to the rotation axis. Unfortunately, the rotation axis inclination i for WR stars is undefined because of their dense winds, making a measurement of the projected rotation velocity $v \sin i$ using broad emission lines impossible. Since the rotation period and the limb-darkening are also unknown for WR 55, Hubrig et al. (2020b) could only estimate a minimum dipole strength of ~1.13 kG using the relation $B_d \geqslant 3\,|\langle B_z \rangle_{\max}|$ (Babcock 1958). Notably, the magnetic field is usually diagnosed in the line-forming regions, which fall fairly far out in a WR wind and not at the stellar surface. Different lines form over different zones of the wind, sampling

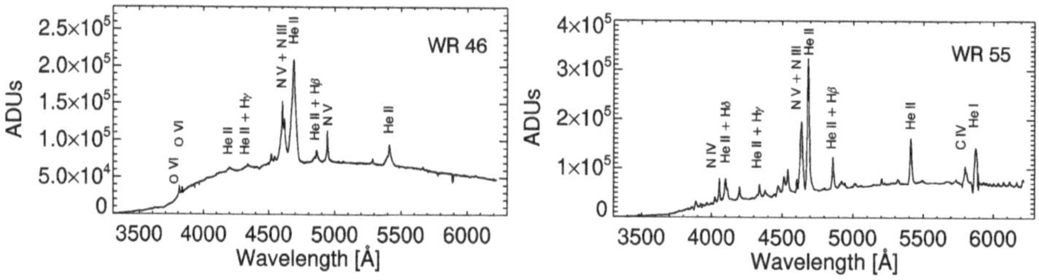

Figure 7.30. FORS 2 Stokes I spectra of WR 46 and WR 55. The spectral line identification is based on the work of Hamann et al. (2006) and is shown above the line profiles. Credit: Figure 1 from Hubrig et al. (2020c).

different field strengths. Thus, the measurements by Hubrig et al. (2020b) for the mean longitudinal magnetic field in WR stars only give a lower limit for the field strength in WR 55: it is expected that the surface magnetic field is significantly stronger than the field measured in the wind lines (de la Chevrotière et al. 2013).

Although magnetic fields are now believed to play an important role in the evolution of massive stars, spectropolarimetric observations of WR stars, which are descendants of massive O stars, are still very scarce. WR stars are usually rather faint: both WR 46 and WR 55 have visual magnitudes $m_V \geqslant 10.9$ and have never been observed spectropolarimetrically in the past. So far, the acquired FORS 2 observations were the first to explore the magnetic nature of these targets. The first detection of the presence of a magnetic field in WR 55 makes this star the best candidate for long-term spectropolarimetric monitoring. Future observing campaigns should be based on spectropolarimetric time series to further strengthen the evidence for the magnetic nature of this star and to set more stringent limits to its magnetic field strength. The temporal variations of the measured longitudinal magnetic field should be used to ascertain the rotation/magnetic period of WR 55 and to determine for the first time the geometry of the global magnetic field in a WR star. Importantly, as WR 55 is associated with the ring nebula RCW 78 and its molecular environment, a study of its magnetic field geometry is of exceptional importance for the understanding of star formation.

Knowledge about the magnetic field geometry in WR stars is furthermore of extreme importance because organised magnetic fields have an impact on stellar wind properties. The interaction of the global magnetic field with the stellar wind was for the first time studied by Küker (2017) using MHD simulations. MHD simulations with an outer boundary at ten stellar radii were carried out for a WR star assuming a dipole strength of 1.1 kG. The disk appeared distinctly fractured and blobs of enhanced density in the disk caused a wake pattern in the wind above the disk. With increasing field strength, the total mass loss rate became more variable, as more and more gas was trapped in the closed field line region and was only released when the field configuration close to the star changed.

References

Abt, H. A., & Moyd, K. I. 1973, ApJ, 182, 809

Abt, H. A., Conti, P. S., Deutsch, A. J., & Wallerstein, G. 1968, ApJ, 153, 177

Abt, H. A., Chaffee, F. H., & Suffolk, G. 1972, ApJ, 175, 779

Abt, H. A. 1961, ApJS, 6, 37

Adelman, S. J., Gulliver, A. F., Kochukhov, O. P., & Ryabchikova, T. A. 2002, ApJ, 575, 449

Alonso, M. S., López-García, Z., Malaroda, S., & Leone, F. 2003, A&A, 402, 331

Arenou, F., Luri, X., Babusiaux, C., et al. 2018, A&A, 616, A17

Arlt, R., & Rüdiger, G. 2011, MNRAS, 412, 107

Arlt, R., Hollerbach, R., & Rüdiger, G. 2003, A&A, 401, 1087

Babcock, H. W. 1958, ApJS, 3, 141

Balona, L. A. 2016, MNRAS, 457, 3724

Barker, P. K., Landstreet, J. D., Marlborough, J. M., & Thompson, I. B. 1985, ApJ, 288, 741

Berghoefer, T. W., Schmitt, J. H. M. M., & Cassinelli, J. P. 1996, A&AS, 118, 481

Blazère, A., Petit, P., Lignières, F., et al. 2016, A&A, 586, A97
Borra, E. F., Landstreet, J. D., & Thompson, I. 1983, ApJS, 53, 151
Breger, M. 1990, DSSN, 2, 21
Briquet, M., Korhonen, H., González, J. F., Hubrig, S., & Hackman, T. 2010, A&A, 511, A71
Briquet, M., Neiner, C., Aerts, C., et al. 2012, MNRAS, 427, 483
Brown, J. C., Telfer, D., Li, Q., et al. 2004, MNRAS, 352, 1061
Budaj, J. 1999, MNRAS, 310, 419
Buysschaert, B., Neiner, C., Briquet, M., & Aerts, C. 2017, A&A, 605, A104
Cantiello, M., & Braithwaite, J. 2011, A&A, 534, A140
Cappa, C. E., Rubio, M., Martín, M. C., & Romero, G. A. 2009, A&A, 508, 759
Carquillat, J. M., & Prieur, J. L. 2007, MNRAS, 380, 1064
Cassinelli, J. P., Brown, J. C., Maheswaran, M., Miller, N. A., & Telfer, D. C. 2002, ApJ, 578, 951
Castelli, F., & Hubrig, S. 2004a, A&A, 425, 263
Castelli, F., & Hubrig, S. 2004b, A&A, 421, L1
Chené, A. N., & St-Louis, N. 2011, ApJ, 736, 140
Chochol, D., Juza, K., Zverko, J., Ziznovsky, J., & Mayer, P. 1988, BAICz, 39, 69
Chojnowski, S. D., Hubrig, S., Hasselquist, S., et al. 2020, MNRAS, 496, 832
Cole, W. A., Fekel, F. C., Hartkopf, W. I., McAlister, H. A., & Tomkin, J. 1992, AJ, 103, 1357
Conti, P. S., & Massey, P. 1989, ApJ, 337, 251
Conti, P. S. 1965, ApJS, 11, 47
Cowley, C. R., Hubrig, S., & González, J. F. 2009, MNRAS, 396, 485
Cowley, C. R., Hubrig, S., Palmeri, P., et al. 2010, MNRAS, 405, 1271
de la Chevrotière, A., St-Louis, N., Moffat, A. F. J., & MiMeS Collaboration, 2013, ApJ, 764, 171
de la Chevrotière, A., St-Louis, N., Moffat, A. F. J., & MiMeS Collaboration, 2014, ApJ, 781, 73
Donati, J. F., Semel, M., & Rees, D. E. 1992, A&A, 265, 669
Donati, J. F., Semel, M., Carter, B. D., Rees, D. E., & Collier Cameron, A. 1997, MNRAS, 291, 658
Ferrario, L., Pringle, J. E., Tout, C. A., & Wickramasinghe, D. T. 2009, MNRAS, 400, L71
Folsom, C. P., Kochukhov, O., Wade, G. A., Silvester, J., & Bagnulo, S. 2010, MNRAS, 407, 2383
Fossati, L., Castro, N., Morel, T., et al. 2015, A&A, 574, A20
Fowler, W. A., Burbidge, E. M., Burbidge, G. R., & Hoyle, F. 1965, ApJ, 142, 423
Freyhammer, L. M., Kurtz, D. W., Elkin, V. G., et al. 2009, MNRAS, 396, 325
Gayley, K. G., & Ignace, R. 2010, ApJ, 708, 615
Gayley, K. G., & Ignace, R. 2012, in AIP Conf. Ser. 1429, Stellar Polarimetry: from Birth to Death, ed. J. L. Hoffman, J. Bjorkman, & B. Whitney (Melville, NY: AIP), 259
Gayley, K. G. 2017, ApJ, 851, 113
González, J. F., & Levato, H. 2006, A&A, 448, 283
González, J. F., Hubrig, S., & Castelli, F. 2010, MNRAS, 402, 2539
González, J. F., Saffe, C., Castelli, F., et al. 2014, A&A, 561, A63
Gosset, E., De Becker, M., Nazé, Y., et al. 2011, A&A, 527, A66
Grevesse, N., Asplund, M., Sauval, A. J., & Scott, P. 2010, Ap&SS, 328, 179
Grunhut, J. H., & Neiner, C. 2015, in IAU Proc. 305, Polarimetry, ed. K. N. Nagendra, S. Bagnulo, R. Centeno, & M. Jesús Martínez González (Cambridge: Cambridge Univ. Press), 53

Grunhut, J. H., Wade, G. A., Neiner, C., et al. 2017, MNRAS, 465, 2432
Guthrie, B. N. G. 1986, MNRAS, 220, 559
Hénault-Brunet, V., St-Louis, N., Marchenko, S. V., et al. 2011, ApJ, 735, 13
Hamann, W. R., Gräfener, G., & Liermann, A. 2006, A&A, 457, 1015
Hamann, W. R., Gräfener, G., Liermann, A., et al. 2019, A&A, 625, A57
Henrichs, H. F., de Jong, J. A., Donati, D. F., et al. 2000, in Magnetic Fields of Chemically Peculiar and Related Stars, ed. Y. V. Glagolevskij, & I. I. Romanyuk, 57
Henrichs, H. F., Neiner, C., Schnerr, R. S., et al. 2009, in IAU Symp. 259, Cosmic Magnetic Fields: From Planets, to Stars and Galaxies, ed. K. G. Strassmeier, A. G. Kosovichev, & J. E. Beckman (Cambridge: Cambridge Univ. Press), 393
Hubrig, S., & Berghofer, T. W. 1998, in IAU Symp. 188, The Hot Universe, ed. K. Koyama, S. Kitamoto, & M. Itoh (Dordrecht: Kluwer), 217
Hubrig, S., & Castelli, F. 2001, A&A, 375, 963
Hubrig, S., & Mathys, G. 1995, ComAp, 18, 167
Hubrig, S., Castelli, F., & Mathys, G. 1999a, A&A, 341, 190
Hubrig, S., Castelli, F., & Wahlgren, G. M. 1999b, A&A, 346, 139
Hubrig, S., Le Mignant, D., North, P., & Krautter, J. 2001, A&A, 372, 152
Hubrig, S., Briquet, M., Schöller, M., et al. 2006a, MNRAS, 369, L61
Hubrig, S., González, J. F., Savanov, I., et al. 2006b, MNRAS, 371, 1953
Hubrig, S., North, P., Schöller, M., & Mathys, G. 2006c, AN, 327, 289
Hubrig, S., Yudin, R. V., Pogodin, M., Schöller, M., & Peters, G. J. 2007, AN, 328, 1133
Hubrig, S., Briquet, M., Morel, T., et al. 2008a, A&A, 488, 287
Hubrig, S., González, J. F., & Arlt, R. 2008b, CoSka, 38, 415
Hubrig, S., Schöller, M., Schnerr, R. S., et al. 2008c, A&A, 490, 793
Hubrig, S., Briquet, M., De Cat, P., et al. 2009a, AN, 330, 317
Hubrig, S., Schöller, M., Savanov, I., et al. 2009b, AN, 330, 708
Hubrig, S., González, J., & Schöller, M. 2010a, in ASP Conf. Ser. 435, Binaries—Key to Comprehension of the Universe, ed. A. Prša, & M. Zejda (San Francisco, CA: ASP), 257
Hubrig, S., Savanov, I., Ilyin, I., et al. 2010b, MNRAS, 408, L61
Hubrig, S., González, J. F., Ilyin, I., et al. 2011a, AN, 332, 998
Hubrig, S., Ilyin, I., Briquet, M., et al. 2011b, A&A, 531, L20
Hubrig, S., Schöller, M., Kharchenko, N. V., et al. 2011c, A&A, 528, A151
Hubrig, S., González, J. F., Ilyin, I., et al. 2012, A&A, 547, A90
Hubrig, S., Schöller, M., Ilyin, I., et al. 2013, A&A, 551, A33
Hubrig, S., Carroll, T. A., Gonzalez, J. F., et al. 2014a, MNRAS, 440, L6
Hubrig, S., Castelli, F., González, J. F., et al. 2014b, MNRAS, 442, 3604
Hubrig, S., Scholz, K., Hamann, W. R., et al. 2016, MNRAS, 458, 3381
Hubrig, S., Ilyin, I., Kholtygin, A. F., Schöller, M., & Skarka, M. 2017, AN, 338, 926
Hubrig, S., Sidoli, L., Postnov, K., et al. 2018, MNRAS, 474, L27
Hubrig, S., Kholtygin, A. F., Sidoli, L., Schöller, M., & Järvinen, S. P. 2019a, in IAU Symp. 346, High-mass X-ray Binaries: Illuminating the Passage from Massive Binaries to Merging Compact Objects (Cambridge: Cambridge Univ. Press), 40
Hubrig, S., Sidoli, L., Postnov, K. A., et al. 2019b, in IAU Symp. 346, High-mass X-ray Binaries: Illuminating the Passage from Massive Binaries to Merging Compact Objects (Cambridge: Cambridge Univ. Press), 193
Hubrig, S., Järvinen, S. P., Korhonen, H., et al. 2020a, MNRAS, 495, L97

Hubrig, S., Schöller, M., Cikota, A., & Järvinen, S. P. 2020b, MNRAS, 499, L116
Hubrig, S. 1998, CoSka, 27, 296
Isobe, S. 1991, PASAu, 9 270
Järvinen, S. P., Hubrig, S., Ilyin, I., Schöller, M., & Briquet, M. 2017, MNRAS, 464, L85
Küker, M. 2017, AN, 338, 868
Kambe, E., Ando, H., Hirata, R., et al. 1993, PASP, 105, 1222
Karitskaya, E. A., Bochkarev, N. G., Hubrig, S., et al. 2010, IBVS, 5950, 1
Kochukhov, O., Johnston, C., Alecian, E., Wade, G. A., & BinaMIcS Collaboration, 2018, MNRAS, 478, 1749
Koenigsberger, G., Moreno, E., & Harrington, D. M. 2012, A&A, 539, A84
Korhonen, H., González, J. F., Briquet, M., et al. 2013, A&A, 553, A27
Kreykenbohm, I., Wilms, J., Kretschmar, P., et al. 2008, A&A, 492, 511
Kupka, F., Dubernet, M. L., & VAMDC Collaboration, 2011, BaltA, 20, 503
Lamontagne, R., Moffat, A. F. J., & Lamarre, A. 1986, AJ, 91, 925
Lanz, T., & Mathys, G. 1993, A&A, 280, 486
Lignières, F., Petit, P., Böhm, T., & Aurière, M. 2009, A&A, 500, L41
Lorenzo, J., Negueruela, I., Castro, N., et al. 2014, A&A, 562, A18
Maheswaran, M., & Cassinelli, J. P. 2009, MNRAS, 394, 415
Maheswaran, M. 2003, ApJ, 592, 1156
Makaganiuk, V., Kochukhov, O., Piskunov, N., et al. 2011, A&A, 525, A97
Makaganiuk, V., Kochukhov, O., Piskunov, N., et al. 2012, A&A, 539, A142
Martínez-Núñez, S., Kretschmar, P., Bozzo, E., et al. 2017, SSRv, 212, 59
Mathys, G., & Hubrig, S. 1995, A&A, 293, 810
Mathys, G., & Hubrig, S. 2006, A&A, 453, 699
Mathys, G., & Lanz, T. 1990, A&A, 230, L21
Mathys, G. 1993, in ASP Conf. Ser. 44, IAU Colloq. 138: Peculiar versus Normal Phenomena in A-type and Related Stars, ed. M. M. Dworetsky, F. Castelli, & R. Faraggiana (San Francisco, CA: ASP), 232
Merrill, P. W., & Burwell, C. G. 1943, ApJ, 98, 153
Michaud, G. 1970, ApJ, 160, 641
Michaud, G., Tarasick, D., Charland, Y., & Pelletier, C. 1983, ApJ, 269, 239
Michaux, Y. J. L., Moffat, A. F. J., Chené, A.-N., & St-Louis, N. 2014, MNRAS, 440, 2
Michielsen, M., Pedersen, M. G., Augustson, K. C., Mathis, S., & Aerts, C. 2019, A&A, 628, A76
Molina, E., & Bosch-Ramon, V. 2020, A&A, 641, A84
Morel, T., Butler, K., Aerts, C., Neiner, C., & Briquet, M. 2006, A&A, 457, 651
Mullan, D. J. 1984, ApJ, 283, 303
Nardetto, N., Mathias, P., Fokin, A., et al. 2013, A&A, 553, A112
Neiner, C., Geers, V. C., Henrichs, H. F., et al. 2003a, A&A, 406, 1019
Neiner, C., Henrichs, H. F., Floquet, M., et al. 2003b, A&A, 411, 565
Neiner, C., Hubert, A. M., Frémat, Y., et al. 2003c, A&A, 409, 275
Niemczura, E., Hümmerich, S., Castelli, F., et al. 2017, NatSR, 7, 5906
Niemczura, E., Kołaczek-Szymański, P. A., Castelli, F., et al. 2020, in IAU Proc. 350, Laboratory Astrophysics: From Observations to Interpretation, ed. F. Salama, & H. Linnartz, 412
Nordstrom, B., & Johansen, K. T. 1994, A&A, 291, 777
North, P., Jaschek, C., Hauck, B., et al. 1997, in ESA Symp. 402, Hipparcos Venice'97, ed. R. M. Bonnet, E. Høg, & P. L. Bernacca, 239

Nuñez, N. E., González, J. F., & Hubrig, S. 2011, in Magnetic Stars, ed. I. I. Romanyuk, & D. O. Kudryavtsev, 361

Oskinova, L. M., Nazé, Y., Todt, H., et al. 2014, NatCo, 5, 4024

Ozuyar, D., Caliskan, S., Stevens, I. R., & Elmasli, A. 2018, PASA, 35, 34

Paunzen, E., Hümmerich, S., & Bernhard, K. 2021, A&A, 645, A34

Persi, P., Fiocchi, M., Tapia, M., et al. 2015, AJ, 150, 21

Petit, P., Lignières, F., Wade, G. A., et al. 2010, A&A, 523, A41

Petit, P., Lignières, F., Aurière, M., et al. 2011, A&A, 532, L13

Piskunov, N. 2008, PhST, 133, 014017

Postnov, K. A., Gornostaev, M. I., Klochkov, D., et al. 2015, MNRAS, 452, 1601

Pourbaix, D., Tokovinin, A. A., Batten, A. H., et al. 2004, A&A, 424, 727

Proffitt, C. R., & Michaud, G. 1989, ApJ, 345, 998

Rachkovskaya, T. M., Lyubimkov, L. S., & Rostopchin, S. I. 2006, ARep, 50, 123

Ramiaramanantsoa, T., Moffat, A. F. J., Chené, A.-N., et al. 2014, MNRAS, 441, 910

Renson, P., & Manfroid, J. 2009, A&A, 498, 961

Rivinius, T., Shultz, M., & Wade, G. A. 2015, in IAU Symp. 307, New Windows on Massive Stars, ed. G. Meynet, C. Georgy, J. Groh, & P. Stee, 228

Roman, N. G., Morgan, W. W., & Eggen, O. J. 1948, ApJ, 107, 107

Romano, P., Sidoli, L., Cusumano, G., et al. 2009, ApJ, 696, 2068

Saesen, S., Briquet, M., & Aerts, C. 2006, CoAst, 147, 109

Savanov, I., & Hubrig, S. 2003, A&A, 410, 299

Savanov, I. S., Hubrig, S., González, J. F., & Schöller, M. 2009, in IAU Symp. 259, Cosmic Magnetic Fields: From Planets, to Stars and Galaxies, ed. K. G. Strassmeier, A. G. Kosovichev, & J. E. Beckman, 401

Savanov, I. S. 1994, Ap, 37, 112

Schöller, M., Correia, S., Hubrig, S., & Ageorges, N. 2010, A&A, 522, A85

Schöller, M., Hubrig, S., Fossati, L., et al. 2017, A&A, 599, A66

Shakura, N. I., & Postnov, K. 2016, in Proc. Accretion Processes in Cosmic Sources, 40

Shakura, N., Postnov, K., Kochetkova, A., & Hjalmarsdotter, L. 2012, MNRAS, 420, 216

Shakura, N. I., Postnov, K. A., Kochetkova, A. Y., et al. 2015, ARep, 59, 645

Shobbrook, R. R. 1985, MNRAS, 212, 591

Shultz, M., Wade, G. A., Rivinius, T., et al. 2017, MNRAS, 471, 2286

Sidoli, L., Paizis, A., & Mereghetti, S. 2006, A&A, 450, L9

Sidoli, L., Romano, P., Mereghetti, S., et al. 2007, A&A, 476, 1307

Sidoli, L. 2017, in XII Multifrequency Behaviour of High Energy Cosmic Sources Workshop, PoS(MULTIF2017)052

Smith, K. C., & Dworetsky, M. M. 1993, A&A, 274, 335

Smith, M. A., Grady, C. A., Peters, G. J., & Feigelson, E. D. 1993, ApJL, 409, L49

Smith, M. A., Lopes de Oliveira, R., & Motch, C. 2016, AdSpR, 58, 782

Smith, M. A. 1994, in IAU Symp. 162, Pulsation, Rotation, and Mass Loss in Early-Type Stars, ed. L. A. Balona, H. F. Henrichs, & J. M. Le Contel (Dordrecht: Kluwer), 241

St-Louis, N., Dalton, M. J., Marchenko, S. V., Moffat, A. F. J., & Willis, A. J. 1995, ApJL, 452, L57

Strassmeier, K. G., Carroll, T. A., & Ilyin, I. V. 2019, A&A, 625, A27

Thompson, T. A., Chang, P., & Quataert, E. 2004, ApJ, 611, 380

Titus, J., & Morgan, W. W. 1940, ApJ, 92, 256

Tokovinin, A. A. 1997, A&AS, 124, 75
Tout, C. A., Wickramasinghe, D. T., Liebert, J., Ferrario, L., & Pringle, J. E. 2008, MNRAS, 387, 897
Tutukov, A. V., & Fedorova, A. V. 2010, ARep, 54, 156
ud-Doula, A., & Owocki, S. P. 2002, ApJ, 576, 413
van den Heuvel, E. P. J., & Heise, J. 1972, NPhS, 239, 67
Veen, P. M., van Genderen, A. M., Crowther, P. A., & van der Hucht, K. A. 2002a, A&A, 385, 600
Veen, P. M., van Genderen, A. M., & van der Hucht, K. A. 2002b, A&A, 385, 619
Wahlgren, G. M., Ilyin, I., & Kochukhov, O. 2001, AAS Meeting 199, 135.04
Watson, W. D. 1971, A&A, 13, 263
Zahn, J. P. 1984, in IAU Symp. 105, Observational Tests of the Stellar Evolution Theory, ed. A. Maeder, & A. Renzini (Dordrecht: Reidel), 379

Swetlana Hubrig and Markus Schöller

Chapter 8

Final Remarks

Major improvements in the studies of stellar magnetic fields in non-degenerate stars with radiative envelopes were achieved in the last decades due to the availability of efficient spectropolarimeters and improved analysis tools. Currently, Ap and Bp stars form the best studied group of early-type magnetic stars. They possess large-scale organised magnetic fields and in almost all of them resembles the predominant component of the magnetic field a dipole whose axis is inclined to the stellar rotation axis. Importantly, these stars represent a unique laboratory to study various physical effects of magnetic fields in stellar atmospheres. The anomalies and the inhomogeneities of chemical abundances observed in the outer atmospheric layers constitute one of the few directly observable manifestations of hydrodynamical processes that occur in the interiors of stars, so that the study of these effects is essential for understanding this fundamentally important but still poorly understood branch of stellar physics. Also answering the question on the origin of their large-scale organised magnetic fields, and in particular the question on the simultaneous existence of non-magnetic A and B and of magnetic Ap and Bp stars, shall have major implications for our understanding of stellar formation and stellar early evolution.

Especially the study of stellar magnetic fields of massive early B- and O-type stars and their role in stellar evolution is of special astrophysical significance, since the presence of magnetic fields is related to phenomena from the era of re-ionisation up to the death of nearby massive stars. Understanding the evolution of high-mass stars remains one of the key questions in astrophysics, as these objects play a major role in shaping the interstellar medium due to their strong UV radiation fields and stellar winds and enrich their environment with heavy elements when exploding as a supernova.

Although significant progress has been achieved in previous surveys aimed at the detection of magnetic fields in massive stars, these surveys usually concentrated on rather bright stars in our Galaxy and did not permit to build trustworthy statistics on

doi:10.1088/2514-3433/abefccch8

the occurrence of magnetic fields in massive stars. No systematic survey of the occurrence of magnetic fields in massive stars at different evolutionary phases and in different environments has been carried out yet. However, trustworthy statistics on the occurrence of magnetic fields is of major importance to identify and model the physical processes responsible for the magnetic field generation inside of these stars. The determination of the strength and structure of the stellar magnetic fields is also fundamental for understanding stellar evolution. Particularly for massive stars with strong winds, magnetic fields need to be taken into account in the calculation of their evolutionary tracks.

The value of spectropolarimetry is generally underestimated in spite of the fact that it permits to probe physical properties in a large variety of objects and which are not otherwise accessible to observations in integral light. A definite shortcoming in the studies of stellar magnetic fields is the paucity of analysing equipment for polarised light at high and low-resolution spectrographs installed at telescopes with large apertures. Thus, the best knowledge of magnetic field strength and geometry is currently limited to bright magnetic stars that were predominantly observed using spectropolarimeters operating on smaller telescopes with apertures between 2 and 3.6 m. As spectropolarimetry requires the collection of a large number of photons, observations using smaller telescopes suffer from long integration times, especially on sources that rotate fast or show the presence of short-term dynamical processes such as pulsations or outbursts. Usually, the polarised flux of most objects does not exceed 1% of the total flux and the polarisation signal should be measured with an accuracy of at least one-tenth. It is also obvious that only spectropolarimeters installed at telescopes with large apertures can be used for surveys of magnetic fields in stars in different environments including galaxies in our neighbourhood. Furthermore, high-resolution spectropolarimeters attached to telescopes with large apertures would permit not only studies of magnetic fields in stars with dynamical atmospheres, but also studies of accurate position and shapes of spectral lines recorded in both circularly and linearly polarised light.

The only high-resolution spectropolarimeter at a telescope with a large aperture is the Potsdam Echelle Polarimetric and Spectroscopic Instrument (PEPSI) attached to the Large Binocular Telescope (LBT) in Arizona and designed to utilise both 8.4 m apertures. PEPSI became a facility instrument in 2020. Almost two decades ago, Donati et al. (2002) suggested to use the high-resolution Ultraviolet and Visual Echelle Spectrograph (UVES) installed at the 8 m Very Large Telescope for magnetic field observations in combination with a Zeeman analyser by building a fibre injection unit into FORS 1 and bring fibres to UVES. This concept avoids reflections under incidence angles far away from zero before the polarisation splitting (as would be the case in a Nasmyth focus) and allows the use of a stable spectrograph not under the influence of changing gravity directions (as would be the case in a Cassegrain focus). Schöller (2019) suggested to build a polarisation injection module for UVES (PIÚ), with a core injection unit similar to that used for PEPSI. The implementation of PIÚ would bring a unique capability to the Very Large Telescope and would permit astronomers for the first time to carry out high-resolution spectropolarimetric observations of faint targets in the Southern hemisphere. While it is desirable for the line of research described in this book to include spectropolarimetric capabilities in the next generation of 30 m-class telescopes, it is not yet clear how this will be incorporated in their highly complex optical designs. Finally, to allow further advances in the field of polarimetry at optical wavelengths, it is of utmost importance that the polarisation capabilities of telescopes are treated as an integral part in their optical design, as is the rule in radio telescopes.

References

Donati, J. F., Catala, C, Mathys, G, et al. 2002, in ESO Astrophysics Symp., Scientific Drivers for ESO Future VLT/VLTI Instrumentation, ed. J Bergeron, & G Monnet (Berlin: Springer), 238

Schöller, M 2019, in The Very Large Telescope in 2030, 27

www.ingramcontent.com/pod-product-compliance
Ingram Content Group UK Ltd.
Pitfield, Milton Keynes, MK11 3LW, UK
UKHW061857100926
8556IPUK00008B/912

9 780750 323901